Finding Middle Ground:
Conversations across America about climate change

By Meera Subramanian

Copyright © 2018 InsideClimate News

All rights reserved.

ISBN: 9781691447800
ISBN-13: 9781691447800

CONTENTS

Introduction	v
In Georgia's Peach Orchards, Warm Winters Raise Specter of Climate Change	1
Seeing God's Hand in the Deadly Floods, Yet Wondering About Climate Change	11
As Snow Disappears, a Family of Dogsled Racers in Wisconsin Can't Agree Why	22
In West Texas Where Wind Power Means Jobs, Climate Talk is Beside the Point	33
Fly-Fishing on Montana's Big Hole River, Signs of Climate Change Are All Around	44
The Flash Drought Brought Misery, but Did It Change Minds on Climate Change?	55
They Know Seas Are Rising, but They're Not Abandoning Their Beloved Cape Cod	66
Generation Climate: Can Young Evangelicals Change the Climate Debate?	77
It's 'Going to End with Me': The Fate of Gulf Fisheries in a Warming World	88
About the Author	102
About Us	103

INTRODUCTION

It's not easy to portray the middle ground on climate change, an issue frequently sketched in black and white as a battle between deniers and alarmists.

As the Trump administration was stepping into the White House in 2017, Meera Subramanian set out for the heart of red America in search of middle ground in Americans' understanding of climate change.

She needed to understand—and convey—the complicated nature of belief, for belief represents so much of what these stories were about: who people trust with knowledge, what constitutes fact. Meera found communities where climate impacts are upending lives. She listened to the people she met, earned their trust and told their stories. First in Musella, Georgia, then in White Sulphur Springs, West Virginia; Gleason, Wisconsin; and Sweetwater, Texas.

But the storyteller in her was not done. In 2018 it was on to Wise River, Montana and Divide County, Montana. And then to Cape Cod, Massachusetts; Wheaton, Illinois; and Palacios Texas. Connect the dots between all those places and you enclose the geographical heart of the country.

In the towns she visited, Meera became a fixture on Main Street, where she spoke with dozens of people. She listened, questioned and listened again as she asked about the changing climate inside diners, gun shops, orchards and churches.

From each place, Meera wrote about the stuff of daily life—peaches and the winter chill, dogs and snow, floodwater and faith, the wind and the future. She examined what happens to people when the world they inhabit suddenly becomes unreliable.

She writes without casting judgment or suggesting blame—right from the middle ground of herself. It's why she can see the middle ground in others, and as you read her stories you can't help but recognize this territory in yourself. Meera's writing is clear, insightful and irresistible. Her storytelling is many-layered. It catches you off guard because it is fundamentally healing, a sorely needed work of understanding worthy of your time.

David Sassoon, Publisher

IN GEORGIA'S PEACH ORCHARDS, WARM WINTERS RAISE SPECTER OF CLIMATE CHANGE

Three generations of Robert Lee Dickeys faced a failed crop after an unusually warm winter. They talk about it as weather rather than climate change.

MUSELLA, Georgia — Three generations of Robert Lee Dickeys share the two chairs in the cozy office of Dickey Farms, the younger always deferring to the elder. For 120 years, the Dickeys have been producing peaches so juicy they demand to be eaten over the kitchen sink.

Robert Lee "Mr. Bob" Dickey II, 89, is slightly stooped but moves quickly, dropping in just for a morning read of the *Wall Street Journal*. His son Robert Dickey III, 63, and his grandson, who goes by Lee, age 33, stick around all day, fielding calls and customers, checking the orchards. The next-generation Dickey is having her morning nap and will appear later in a tiny flowered dress, cradled in the arms of her mother, Lee's wife, Stacy.

Just outside the office is the retail shop, where I watch customers drift into an open-air porch with white rocking chairs and a breeze, to consider peaches. Or, rather, the lack of peaches.

It's mid-July, what should be peak season, but the only variety on offer is Zee Ladies, almost the last of this year's fruit. Behind the cash registers, the peach production line is still and silent, lights switched off.

In a normal year, midsummer would be abuzz with workers packing July Prince peaches in boxes they pull from hooks swirling overhead. But this year, about 85 percent of Georgia's peach crop failed. It wasn't a freeze, though they did lose some fruit to a mid-March dip into the 20s. And it wasn't hail, though a hail storm in early April took some, too. The harvest failed because it was a warm winter. A very warm winter, even warmer than the warm winter the year before.

It was 1990 when, sitting in an undergraduate biology class at the University of Georgia in Athens, I first heard the term "global warming." I remember only one fact the professor offered that day: if the Earth's temperature continued its apparent rise, peaches would no longer be able to grow in the Peach State of Georgia. Now, 27 years later, it was looking like that prophecy was coming true. Could this year's ruined crop be a harbinger of warmer winters to come?

"I was very skeptical two years ago," Mr. Bob's son Robert says. "But with two warm winters I'm beginning to pay a lot more notice to it."

I ask how many consecutive winters he'd have to experience before he started planting varieties that could handle warmer weather. He laughs, then says, "Maybe one more."

Naming the Problem

An iconic sweet Georgia peach might be the hallmark of summer, but its life cycle begins in darkest winter, deep inside still-bare tree branches. Most winters, when cold fronts are dumping snow up north, frigid air sweeps down South, causing something physiological to happen in the cells of buds that will send forth new leaves and bear the stone fruit come spring. It's the cyclical reset button, the seeming nothingness that allows everything else to occur.

Peaches—along with many other fruits and nuts, from apples to walnuts—need cold like you need sleep, not just any sleep but dream-state sleep, the deeper and more sustained the better. This year, they did not get it.

The Dickeys have been peach farmers since 1897, when Mr. Bob's grandfather first planted trees in the dirt of middle Georgia, where the soil and elevation and water serve the crop well. Along with a handful of nearby peach growing operations, the Dickeys now dominate the Georgia peach market, the country's third largest after California and South Carolina. They cultivate a thousand acres of peaches, and nothing but peaches.

"This is one of the few years that I can remember that we didn't have enough cold weather," Mr. Bob tells me as we sit knee-to-knee so he can hear me. "Most of us peach growers, we worry more about spring frost ... but this year the crop was decimated on account of lack of cold weather."

I've come to middle Georgia curious if peach growers were experiencing a changing climate that threatened the state fruit—not to mention their generational legacy—but Mr. Bob insisted this year's crop failure had nothing to do with that thing the politicians call climate change.

Weather, Mr. Bob tells me, "it comes and goes, and you have cycles. I think we're in a warming cycle this year." He looks at me with pale blue eyes and smiles. "It might be cold as mischief next year!" he says, laughing. Soon he is out the door, on the way to a friend's funeral.

The Chill Is Gone

Each year, on the first of October, the Dickeys and other peach growers start counting every hour that dips below 45 degrees Fahrenheit, a "chill hour." Most peaches grown in Georgia need at least 650 of these chill hours, which usually isn't a problem in a region with an historic average of 1,100 chill hours per winter.

But last winter the chill-hour count was only about 450 by the middle of January. Then it just stalled. By Valentine's Day, when the farmers are usually long done counting, the figure hadn't reached 500.

Like all peach farmers, the Dickeys tread a fine line, attempting to grow the peaches that ripen earliest, so they can lead the national market, while not getting caught by late freezes. In early 2017, Georgia's struggle with its lack of chill hours meant a sluggish bloom. The canopy of pink was diminished, the iridescent green of new leaves hampered, even as the days got longer and spring progressed. The trees had not slept, so they did not know to awaken.

This meant that when a March freeze hit, there was almost no damage in Georgia; the trees had barely bloomed. In South Carolina, though, the same freeze was devastating. The South Carolina winter had been just cold enough to adequately stimulate buds, but so warm that they erupted early. With the March freeze, South Carolina farmers lost nearly everything.

According to NOAA, Georgia and South Carolina together suffered $1 billion in peach crop losses this year.

The prior winter had been warm as well, Robert tells me. That year, the peaches pulled through. This year, they did not.

"We hold our breath every year with weather," Robert says. "Chill hours or late freeze, dry weather or hail storm. All kinds of things can get thrown at you in this business."

But warm winters are especially vexing. When crops are lost to freezes and hail storms, the damage is immediate and obvious. The effects of a low-chill year vary wildly among varieties of peaches, between micro-climates within a single orchard, even from the base of the branch to its tip.

"It was very unusual," says Robert, recalling how baffled he was by his orchards this spring. "We did not know what some of the peaches on the tree would do." The sales team made promises they ended up unable to keep. And which 15 percent of your customers do you keep happy?

Just about the only thing peach farmers can do nowadays to try to make up missing chill hours is douse their trees with hydrogen cyanamide. This toxic growth regulator, commonly sold under the brand name Dormex, is thought to simulate chill hours. Growers selectively sprayed this year and got mixed results in terms of fruit, though it did seem to spur leaf growth for tree health. "It can help," another peach grower, Lawton Pearson of Pearson Farms, told me, "but it's not going to pull your ox out of the ditch."

Dormex might address the immediate problem, but looking long-term raises larger questions. When do you begin to consider planting varieties of peaches that need fewer chill hours? When do you decide that Mr. Bob's great-granddaughter will slip on her boots to go work an entirely different crop, one that might come to define Georgia in the 21st century?

'This Weather Thing'

Farmers inhabit the world of weather. "We follow the weather tremendously," Robert says.

Weather is not climate, as climate scientists say again and again. Weather is what you wear on a particular day; climate is your whole wardrobe. But people live day-by-day. They don't go out and buy a wardrobe; they stand before an open bureau drawer on a single morning and decide which shirt to slip on. Or which variety of peach tree to plant.

As much as they care about the weather—and depend on the climate—most of the Georgia peach farmers I met recoiled at the phrase "climate change." I found myself reluctant to reveal that I was writing for *InsideClimate News*, and I listened to farmers who danced around the language as much as I did.

"I'm not all agreeing with this weather thing," Mr. Bob said.

Whether they agree with it or not, though, the climate in their region is getting warmer. NOAA reported that this past winter was the fifth warmest in the eastern United States since record-keeping began 123 years ago, which is also when many of the peach farms of Georgia were just getting started. The peach growers point to other recent data to support their skepticism. Yes, they say, the previous two winters might have been warmer than usual—but the two winters before that were unusually cold. And the weather station in nearby Macon shows that while last winter was the fourth-warmest since 1948, it was outranked by a string of warm winters in the 1940s and '50s—strangely warm winters that growers refer to when confronted with the specter of climate change. They've seen it before, the growers argue, and they'll see it again.

"Sometimes it comes in cycles. We don't really talk about long-term weather changes," Robert says. "We haven't seen that yet, with peaches. Hopefully next year there will be enough chill."

But the earth has been keeping accounts exponentially longer than humans. To a geologist, last century and the next are as good as the present. It's only by peering through the long lens of paleoclimatic history that scientists get a clear picture of the scale of changes underway. They search for evidence amid the deep layers of planetary history, in the remains of glacial moraines and ice core samples.

And what they find is that there hasn't been this much carbon dioxide in the atmosphere for about 3 million years, when it was so warm that oceans reached far inland compared to today and the world was a fundamentally different place.

Frustrations of Farming

In addition to being a fourth-generation peach farmer, Robert is also a second-term state representative, Republican, District 140.

We climb into his Chevy Tahoe, parked between the farm office and a shuttered cotton gin, across the street from the postcard-perfect brick general store his great-grandfather built, and head to his fields. As we drive from orchard to orchard, looking at trees that are green but bereft of fruit, he speaks a lament for rural Georgia. The environment that concerns him most is the business environment.

"Agriculture doesn't have that large employment component that it used to," he says, "so we're looking for manufacturing jobs in the rural part of Georgia. But we don't have the skilled workforce," and companies won't come unless they see that skill.

"We're spending lots of money on technical education. Try to keep these kids in high school from dropping out. I could go on and on, but it's very frustrating. Your better students are flocking to where the jobs are, which is in metro areas." Georgia is made up of small counties. The majority of them (including Musella's Crawford County) have been shrinking in population, Robert said, while the areas that grow have been primarily urban.

Robert's own son Lee left the farm years ago to become a CPA and work finance for big firms in Atlanta. He relinquished a lucrative job last year to return to the family business—only to have his first peach harvest fail.

Generally, the people still living in towns like Musella are what Robert calls "the poor, unskilled and unemployed, and the elderly people." And they don't want the jobs Dickey Farms can offer. "I don't want to browbeat good American people," he says, but "American people aren't gonna do outside hard labor. They'll drive a tractor all day long or sit in the packing house and run a piece of equipment or drive a truck. But get out here and pick peaches or cut limbs, they just won't do it."

We pull into an orchard of Sun Prince peach trees, where Robert consults with a crew leader who's been with Dickey Farms for 28 years. Originally from Mexico and now an American citizen, the crew leader is like one of the family, Robert tells me. The workers, who are on H-2A visas, speak Spanish as they push their collective body weight into moving a massive stack of pruned peach limbs, their work shirts solid with sweat in the midday heat.

Shark's Tooth Prophecy

Even if average temperatures tick up by only one degree, the increasing fluctuations between extreme highs and lows that many climate models predict could gravely affect farmers of all crops. A recent study in the journal *Science* concluded that the Southeast will be hardest hit by climate change on multiple fronts.

"We figure the insect problems are going to be bad" if the winters get warmer, says Jeff Cook, the University of Georgia agricultural extension agent who serves the counties where the Dickeys have orchards. He has been called a "guru" to the local peach growers. "We're worried about diseases getting worse, because a lot of things didn't go completely dormant," he says. But he easily recognizes how the growers could see the upside, too. "I think they'd just say we have a longer growing season. We can grow more stuff."

Meeting three generations of Robert Lees in the Deep South is a reminder that the past is never quite past in middle Georgia. But the past, they insist, does not predict the future. The Dickeys use the word "optimistic." They use the word "hope."

Quietly, they are adapting, innovating, recovering and hedging their bets. They are planting test plots of low-chill peaches and watching them closely. Peaches are immigrants to Georgia from their faraway origins in China. They have always been on the move because of humans, though now human influences could be moving the crop in a less direct yet perhaps more profound way.

The Georgia peach farmers might not like the term "climate change," but in a sense, the evidence of long-term change is right under their feet. While the peach is the state fruit, Georgia also has a state fossil: the shark's tooth. The oceanic predators' toothy remains can be found not just on the coast, but also more than 100 miles inland.

I stood with Jeff Cook in one of the Dickey peach orchards after we'd driven over the hump of Rich Hill, where the road rises 150 feet from the surrounding valley floor. He told me we stood on the fall line that cuts through Georgia, dividing the Piedmont that reaches toward the mountains from the coastal plain that once submerged half of Georgia under an ancient ocean. And he told me stories he'd heard from older farmers, of how amid the striated layers of red sand and limestone that lay bare the collision of two landscapes, they have unearthed shark's teeth from the dirt.

As I listened to Cook, I imagined the peach farmer's plight as a man standing on the fall line clutching two stones: the tooth-turned-to-stone in one hand and the stone fruit of a peach in the other. In my imagination, the two stones were separated only by the steady tick of time and change moving at an unknown pace.

SEEING GOD'S HAND IN THE DEADLY FLOODS, YET WONDERING ABOUT CLIMATE CHANGE

An evangelical mountain town lost eight people to flooding from an extreme rain storm. Many residents see the Biblical prophecy of the apocalypse, and welcome it.

An evangelical mountain town lost eight people to flooding from an extreme rain storm. Many residents see the Biblical prophecy of the apocalypse, and welcome it.

Jake Dowdy is a police officer in White Sulphur Springs, West Virginia, where he lived a block from Howard Creek, a stream so inconsequential you could usually hop-skip across parts of it without wetting your toes.

It was the morning of June 23, 2016, and a heavy rain was falling as Jake went to the gym for a workout. He wasn't thinking much about the rain, other than that it'd be good for the garden. When he got home around noon, he had lunch and kicked up his feet in the living room, chilling out for a while before his 4 pm shift. He drifted off to sleep on the couch and awoke when his wife texted, confusing him for a moment; she was concerned about reports of flooding.

His disorientation turned to panic when he set his feet on the carpet and felt it squish soggily beneath his soles. He had just enough time to grab the cat and wade through thigh-high rushing water to his truck.

Meanwhile Jake's neighbor, Kathy Glover, was at her office job on Main Street. She was aware of the heavy rain but wasn't concerned about the safety of her house, the home she'd lived in since she was 2 years old. It was two long blocks from Howard Creek. But in the early afternoon, a neighbor called to tell her that water was lapping at her front steps and she ought to get home. It wasn't coming up from the creek, but pouring down from Greenbrier Mountain.

Kathy rushed home, and for hours she tried to fend off the water from her front door, until she realized that the creek behind her had risen so much and so fast that it was entering her kitchen through the back door. The house was surrounded. Kathy and her daughter fled for higher ground.

More than a year later, Kathy has rebuilt on the same spot. Jake and his family moved to higher ground; they had no interest in living so close to the creek's edge again. Now there's nothing left in the spot where their house once stood except two small "for sale" signs.

While they differ in how safe they felt returning to their flooded property, Kathy and Jake both speak of the White Sulphur Springs flood, along with other extreme weather events around the world, in terms of the Biblical prophecy of the apocalypse. Their religious reasoning stands in stark contrast to what climate scientists offer as explanation for record-breaking rain storms, but could anything reconcile the divergent views?

A Tale of Two Tellings

Most of those who chose to rebuild along the now-quiet creek are settled into their houses, a few towering on 10-foot foundations while others just meet FEMA's minimum requirement for houses in a flood plain: 2 feet above the 100-year-flood level. Other storm victims, like Kathy, who is officially outside the flood plain, rebuilt their houses just as they were before.

Between the renovated homes on the opposite side of Howard Creek are freshly seeded parks bursting with colorful flowers that memorialize the dead: The three members of the Nicely family who took refuge in their attic, only to have the house tear from the foundation and float away. Mykala Phillips, the 14-year-old girl whose father tried to save her with an extension cord that wasn't strong enough to serve as a lifeline; her body was found months later, miles down the creek and in the next town. Belinda Scott, who was catapulted from her home into a tree after a gas explosion, making it to the hospital but with injuries too severe to survive.

Eight people died in the town of 2,400, and there were 23 total fatalities across the state, with hundreds of houses damaged.

Human beings are natural storytellers, and there are two narratives being told in America today about how to make sense of extreme weather events like the one that swept through White Sulphur Springs in 2016.

One narrative comes from scientists and large international organizations with unwieldy names like the Intergovernmental Panel on Climate Change (IPCC), who offer a lot of numbers and complicated science about a warming planet, right down to the dueling temperature markers of Fahrenheit and Celsius. There are also some uncontroversial bits of data based on the physical fact that warmer air holds more water.

Over the last half-century, this narrative tells us, there has been a 70 percent increase in the amount of precipitation in the heaviest Northeastern storms, from West Virginia up to Maine.

The isolated band of thunderstorms on the day of the White Sulphur Springs flood, which drenched the town with wave after wave of downpours, ultimately dropped a record-breaking 9 inches of rain, twice as much as the previous record rainfall in 1890. According to Dr. Kevin Law, West Virginia's climatologist, it was definitely one of the worst storms in the state's history.

The other narrative turns to a single book filled with parables of good versus evil and drama far more compelling than what the IPCC reports offer. The Bible explains tragedy and human suffering and redemption as being part of God's plan, giving meaning to natural disasters.

"I'm a firm believer that God tells us in the Bible that he will warn us through signs in the sky," Kathy told me when I visited her earlier this fall. "It's fitting in with the Book of Revelation. With the earthquakes and the devastation happening around the world, it's a wake-up call."

Sixty years old, Kathy has blond hair styled short and an easy smile that brings out her dimples. In addition to her job as office manager at Workers United Local 863, she sits on the city council as the recorder.

The flood is "a sign of the times," she said, and feels that her primary duty since the storm is to "share God's word with more people." In terms of actions to be taken in a post-flood town, "There's nothing really I can encourage or discourage, other than to encourage people to be ready for the Return."

A Time of Puzzlement

Two weeks before I arrived in West Virginia, Hurricane Harvey pummeled the Texas Gulf Coast with rainfall that in some spots doubled prior records—more than four feet of rain over four days. And while I was there, Hurricane Irma broke records for sustained wind speeds as she spun disaster over the Caribbean and Florida. Hurricane Maria was still brewing, her devastation yet to come.

These events dominated news headlines, as epic stories tend to. But small-scale flash floods like the one in White Sulphur Springs accounted for the second-highest number of weather-related fatalities in 2016; only heat killed more people. These sudden floods hit Houston in April of that year, Maryland in July, and Louisiana in August, where they took out 40,000 houses. The damage from the rains was worsened by lax building codes and excessive development in vulnerable areas, but the ultimate source of the destruction was the sheer amount of water itself.

Surveying the damage in their own town, White Sulphur Springs residents like Kathy, Jake and many of their friends and neighbors turned to Christianity for an explanation, and for guidance about what to do next.

"I'm a believer in the Good Lord," Jake told me as we sat together in City Hall. At 31 years old, he'd recently been promoted to chief of police and was cognizant of his increased responsibilities in safeguarding the town's citizens.

In full uniform, with his strawberry hair shorn close, he was composed and steady as he described his disorienting experience of the flood, but his demeanor shifted as the conversation veered toward climate change. He seemed genuinely confounded, wrestling with how to square church teachings with recent weather extremes.

"With the flooding not only here but other places, and the wildfires out west, and hurricanes hitting the Eastern Seaboard," he said, stretching his arm over his head, his foot starting to tap-tap below the table, "and you look up in the north and you're seeing glaciers melting—seems like more and more things are impacting this world that are hard to explain. Things are changing. I don't know," he said, "It's kinda scary in a way. I don't know if these are signs of the ending coming or if this is climate change. I'm as puzzled as everybody is."

Who Has the Authority?

Before the flood, many of the people in White Sulphur Springs didn't think about climate one way or the other. In its aftermath, many turned to their churches not only for how to make sense of what had turned their tame creek into a torrent, but for physical relief as well.

Evangelical Christians make up 40 percent of the population of West Virginia (compared to 25 percent nationwide). In the small town of White Sulphur Springs, there are more than 20 churches. Religion permeates many spheres of life: when I asked people in town what they thought about the science of climate change, most punted the question, often to their pastors, who carry an authority not granted to scientists. It's not that people were skeptical of the science necessarily; instead, they seemed to think of it as cordoned off in a place they didn't consider theirs.

"I don't have a big science knowledge or that sort of thing," said Kathy, who attends one of the Baptist churches in town, when I asked her how she thought about the flood.

Chad Dingess, the pastor of the fast-growing evangelical Bethesda Church, said something similar. He could speak of the end of the world, but it was not his place to consider climate change.

"I don't feel like I have enough knowledge to speak on that," Dingess told me early one Sunday morning, before stepping on stage to preach three sermons that would reach one thousand people by noon. He said he has "no clue" if humans have had an impact on climate change.

"The scriptures say very clearly: in the last days, there would be earthquakes and hurricanes. And what the scripture really says is that it will happen more frequently. And I think that's some of what we're seeing," he said. "I don't think it's anything we need to be afraid of. It's pointing to the return of Christ."

When the waves of thunderstorms drenched White Sulphur, believers and non-believers alike were unified in horror that the flood came so fast. Together as a town, they still suffer jitters each time a hard rain falls. But the deeply religious had a broader context that transferred this confusing data into clear "signs." From Matthew 24 to the Book of Revelation, they were primed for apocalypse.

Just as Rachel Carson amalgamated every dire outcome that chemicals could produce to make a point in her opening chapter of *Silent Spring*, the Book of Revelation mentions almost every catastrophic natural tragedy that can befall humanity: locusts and earthquakes, eclipses and scorpion strikes, scorched earth and sulfury fires.

No matter how fast Irma's winds roil or how many feet of rain Harvey dumps, it is a story the faithful have heard before, and—this is the tricky part—in some ways, long for. The greater the collection of disasters, the closer the long-awaited return of Jesus Christ.

Faithful Action

How soon that return might occur has vexed Christianity for a couple of millennia. It's also forced Christians to answer questions about how to pace their lives and where to direct their energy while here. Some focus on soul work, but others are answering with the body work of community action, sometimes specifically related to weather events.

In the aftermath of disasters that the scientifically minded explain by climate change, it's the faith-based groups that show up, many (but not all) of them Christian. Three-quarters of the organizations associated with the National Voluntary Organizations Active in Disaster are faith-based. FEMA might be there writing checks, but that requires a lot of cumbersome paperwork and complicated rules. The Mennonite Disaster Service comes, builds a house, and leaves, refusing even a thank you.

Kathy is grateful for this kind of service; it is part of what allowed her to stretch $13,000 FEMA assistance into the $34,000 she actually needed to rebuild her house. When talking about that feat, Kathy's daughter summons the Biblical story of the few loaves and fishes that miraculously stretched to feed thousands.

"We are where we are today because of God's people that were here on the ground within the first 24 hours, who were helping with the recovery, helping feeding people," Kathy said. "You don't see the government still here, trying to help people rebuild."

In fact, the government *was* still there. A team of FEMA and National Guardsmen at that very moment was overseeing the steady demolition of houses beyond repair more than a year after the flood, and part of Kathy's rebuilding miracle was performed by the secular United Way, which bought all her kitchen appliances.

But it wasn't as impressive to her as the help that came from local churches or the "orange shirts" of Samaritan's Purse, an organization led by the Reverend Franklin Graham. Their actions were easier to spot and appreciate, like the volunteers I saw ushering a family in a nearby town also hit by the flood out of their blocky, government-issue FEMA trailer and into a beautiful, new (elevated) house built by Samaritan's Purse contributions. The young son slid giddily across the shiny wood floor in his socks.

"You see the ministries of the Samaritan's Purse and from the Southern Baptist Convention helping people rebuild," Kathy continued with awe. "They're the people that are still here.

It is God's love expressed through human love, immediate and compassionate, arriving with peanut butter sandwiches when people are still wading out of their homes to high ground; it is God's love expressed through human love that's still there 15 months later, swinging hammers. I heard this from Kathy and many others. There is no denying the generosity, the unity, the vigor that rebuilding inspires. You can see it in the T-shirt slogans: West Virginia Strong. Houston Strong. Fill-in-the-blank Strong. Rally cries of resilience. The absolute best demonstrations of the human spirit.

There will always be weather disasters that will spark the emergence of these heroes. There will always be room for sweet sympathies for storm victims, the outpouring of donations, the arrival of strangers ready to shovel mud out of houses and hand out blankets.

But the other storyline, the scientific one, proffers the possibility of prevention. Can we halt the warming of the Earth and the seas by ratcheting back the amount of greenhouse gases we produce, so less water rises up to later fall as rain, heavier and harder than we've experienced before? So NOAA doesn't have to come up with new colors on rain charts that signal feet rather than inches? So a good heavy rain is really just good for the garden, and you can wake from your afternoon nap and head to a day at work that will sift into a blur of other indistinguishable rainy days?

"I definitely feel that climate change is important," Jake said as we chatted in City Hall. "We need to be taking more precautions and steps to ensure that we're not polluting the planet as much." He paused, and then went into police chief mode. "We'd look at hybrid cruisers," he said with a laugh, "if they were substantial enough for the work that we do."

There is another narrative to be found in the story of scripture. But it doesn't turn to the burning pages of Revelation at the end of the Good Book and what has been called "escapism theology," with a hyper focus on the world to come.

Instead, it highlights the origin story of Genesis at the book's beginning, when God created a world, this world, teeming with living creatures and birds flying in an expansive sky. There was a garden with rivers flowing from it, and the humans God had crafted were placed there to work it and take care of it.

Science is deepening our knowledge of the garden that is our planet, with scientists seeking understanding as urgently as pastors pouring over parables in the Bible, both hoping that come Sunday morning, listeners are ready to receive the message.

AS SNOW DISAPPEARS, A FAMILY OF DOGSLED RACERS IN WISCONSIN CAN'T AGREE WHY

A father and daughter have been running sled dogs for more than 25 years. It's easier for them to talk dogs than politics, weather than climate.

As dog musher Mel Omernick slipped nylon harnesses over her Alaskan huskies' lithe bodies, the dogs were already straining with forward momentum. Pogo pressed her paws into the ground below, the sound of her yelps joining with those of the three other dogs that Mel and her husband, Keith, were hooking up to their tuglines. The cries melded with the barking of a hundred other dogs at the Redpaws Dirty Dog Dryland Derby in northern Wisconsin.

It was the first weekend of November, and race participants had come from all over Wisconsin and neighboring states, and as far away as New Hampshire and Quebec, to run their dogs. All year, they had fed and watered and trained and cleaned up after their teams, awaiting the moment they could let their dogs loose across the starting line.

Now the race weekend had finally arrived, though it had gotten off to a rocky start. Once again, the weather was to blame.

Northern Wisconsin is still a frigid place come winter. But as the state has warmed, the certainty of snow gradually vanished, leaving the traditional winter dogsledding races frequently cancelled for lack of good powder. Organizers responded by adapting the sport itself, from dogsledding to "dryland" racing.

The Dirty Dog Derby was the first of its kind in the area, started in 2006 to extend the racing season into spring and fall so that mushers like Mel and Keith could have more chances to compete, and dogs like Pogo more chances to run. Swapping out sleds, dogs instead pull mushers on unmotorized rigs or a cart with four to 10 dogs or modified bicycles (bikejoring) pulled by two dogs: in some cases, a single musher simply lashes herself by bungee cord to a single dog and runs behind him in an event called canicross. Dryland variations tend to be shorter events, sprints of a few miles instead of the hundreds of miles of the iconic long-distance sled races often associated with the sport.

In the weeks before their race, the Dirty Dog organizers had been worried they'd have to cancel it if the warm weather they were experiencing into late October—still hitting 70 on some days—continued. Since dogs can't sweat—the only means they have to release heat from their bodies is through their tongues and the pads of their paws—mushers won't run their dogs if there's a risk they'll get "fried" by overheating.

But by the time the derby arrived on Nov. 4, race organizers were pining for a little heat. The race grounds—and the carefully groomed trails—were blanketed in nearly three inches of snow. The day's races were cancelled. Mushers kept their spirits up but weren't finding much humor in the irony: a dryland race, the sport's creative solution to a paucity of snowfall, cancelled because of snow. Some mushers loaded up their trailers with their pent-up dogs and made their way back home, while others—usually those who had traveled greater distances—hung around, eating chili from the Dirty Dog Diner set up in the open-air lodge and taking shifts by the fireplace as they waited to see if the weather would change.

By the second day, the snow had melted just enough to turn the trails into a muddy, but navigable, quagmire. When the organizers announced early Sunday morning that the race was on, the grounds erupted in excitement and movement. Mel and Keith headed to their truck to get the team hooked up, and soon the dogs were pulling on their lines, amped to do what they seemed born to do: run.

Dogsledding, Without the Sleds

Today, dogsledding is undergoing a transformation. Or threat, depending on your viewpoint.

The first hit came with the advent of snowmobiles in the 1960s, when dogsledding began slipping away as the standard form of transportation for many of the world's northernmost inhabitants. Instead it became recreational, one of those activities that meld sport, hobby and lifestyle into one expensive, obsessive pastime.

Now the ideal image of dogs, humans and a sled careening silently across snow is facing a new challenge as the climate warms and the weather weirds. The Iditarod, a thousand-mile race across Alaska that is the most famous of sled dog races, had to be rerouted two of the last three years as its organizers chased snow-covered terrain. In Wisconsin, since 2001, about one-third of the sled races failed to happen, primarily because of lack of snow.

"I definitely see a trend where things are not like they used to be," said Jan Bootz-Dittmar, a champion sprint musher on snow and dry land who's been running dogs for 40 years. Last year, insufficient snowfall caused half of the snow races in Wisconsin to be cancelled.

"That affects me," she said in the cafe as she munched on potato chips in lieu of lunch, "and it pisses me off."

The Accidental Life

Mel races drylands, but skijoring is what she loves most: the quiet "shwooosh shwooosh" of her skis gliding through a snow-silenced world but for the sound of her dog's movement.

She lives in Lincoln County, in north-central Wisconsin, and we were talking in the kitchen of the home she and Keith share, a long green metal building divided into a utilitarian shop and a capacious, wood-ceilinged living space with a wall of windows looking out upon a stand of trees. Her parents' home lies just beyond. Mel, 40, sees both of her parents nearly daily, but when it comes to dogsledding, she's closer to her father, Ron Behm, who is approaching 70. The kennel of Alaskan huskies and hounds that was once Ron's is now cared for by Mel and Keith, and they all train the dogs together.

Father and daughter have been running sled dogs for more than 25 years, since they entered the sport by an accident of canine lust. Mel was still in junior high when the neighbor's Malamute wandered over and found their Labrador mix, which was Ron's hunting dog. One lone pup was the result, and they kept him and named him Tiny. He was no bird dog, but Mel and her three siblings kept him running all the time. When the neighbor's dog got loose again the next year and a litter of four was born, the Behm kids had a team. The family was friends with Jan Bootz-Dittmar, who gave them some harnesses to try out.

Mel and her two younger brothers hooked up Tiny and the team to their red Radio Flyer wagon, and the boys would take turns riding while Mel, who was a gymnast at the time and wanted the exercise, darted in front, leading the dog. As the dogs grew, they swapped out the wagon with an old lawn mower, engine removed. Mel's older sister Ginnie, daunted by the speed, would cheer them on, snapping photographs. Ron mowed a path through the grass so the children could holler "gee!" for right and "haw!" for left as the dogs learned commands.

Mel's brother Adam was the first to enter a formal dogsledding race, with Ron joining him a few years later. By the time she was in college, Mel had quit gymnastics and started racing, too. Mel's mother, Gail, stitched harnesses and kept the mushers supplied with baked goods and the fresh perch she caught while ice fishing, her preferred sport.

Dynamics of Differing

Despite all this family togetherness, there was one crucial split in the Behm household: politics.

At first, Mel told me, she was a Republican because her father was. "I didn't pay attention to politics," she said. But that changed when she became an emergency room nurse. Working 12-hour shifts with people in crisis, she suddenly realized that "some of the decisions that politicians were making were affecting my patients."

She also saw how not just politics but also science affected them, from the medicines she could offer them to how their bodies responded. She saw science in her sport, too, where the principles of genetics were used to breed dogs for speed, endurance and tougher paws.

"I love science," she said, "and I believe in evolution." Evolution was one of the science-based subjects she'd argued with her dad about most fiercely when she was in college. "I feel like we're an example of it, and our sled dogs are too. So, it's logical. It just makes sense ... our planet is changing."

These were not conversations Mel and Ron had easily, or often. Usually they just avoided politics—and science—altogether, focusing on the thing that bound them, their love of dogs and dogsledding, their family life.

Within the contained world of dog mushers, there's a similar hesitation to talk about potentially divisive issues. The entire political spectrum finds representation at the Dirty Dog Dryland Derby, from the young women showing up in a Prius with two dogs tucked in back, to the Trump supporter in a trailer emblazoned with "To the victor the spoils." But as with Mel and her father, conversations among the mushers veer away from the political, to the point that many mushers don't even know each other's leanings or affiliations. Better to talk dogs than politics, and weather before climate. Even Ron, once I pressed him, was adamant that he was a "constitutionalist," not a Republican. That was a distinction even his own daughter didn't know.

Just a Blade of Grass

The second weekend in November, a second dryland derby was scheduled. Ron was slated to be a race marshal for that one, known as the Willow Springs Round Barn Fall Rally, and Keith was scheduled to compete; Mel planned to swing by after an all-night ER shift. But the skies stayed heavy most of the week, and, looking at the forecast of more snow, organizers cancelled the race by Tuesday night.

With his weekend freed up, Ron was willing to continue a conversation we'd started earlier in the week about what was happening to the climate and the sport he and his daughter both loved. Instead of race marshaling, he joined me at Mel's house, laying down his coyote fur cap upon the kitchen table as Mel fixed coffee after her 12-hour shift.

Sitting side by side, Mel and Ron are two generations delighted in the world. Ron is still fit from his 30 years as a mail carrier, from which he retired in 2012; he now devotes himself to the Wisconsin Trailblazers race dog club, the Lions and a one-acre market garden that he tends with his wife. His white beard is trimmed, and Mel's blonde hair is cut in a bob that falls in soft curls. They both default to easy smiles, even when their viewpoints clash.

Which they do, when it comes to climate change.

Mel feels that the winters of her youth are gone. Where was ice skating at Thanksgiving, like she remembers from her grandma's when she was a kid?

"It always seemed harsh in the winter," she said. But Ron had an explanation that had nothing to do with climate change. A trick of perception, he said as Mel listened respectfully, since in those days there was none of the high-tech clothing and efficient snow plows of today. He likened it to other mythic stories about one's childhood, à la walking to school, uphill, both ways. "Probably that was a part of it," Mel responded, nodding thoughtfully, both of them disagreeing in a way that was unfailingly polite.

There are 19 dogs out in the kennel, but six are allowed in the house, and they periodically came up to Mel and Ron, who stroked their heads. One dog took a brief interest in Ron's coyote-skin hat—it's roadkill, Ron told me—before venturing off again.

"One thing about weather," Ron said, "we can all comment about it, but we can't change it." He sees climate changes as cyclical, pointing to the fact that long before humans were contributing any sort of emissions to the atmosphere, the state had "gone through three major warming trends, and also, three major freezing trends." He mentioned the nearby Ice Age Trail that marks the edge of the last glaciation, 10,000 years ago.

But even as he referred to deep time and geological history, Ron expressed his strong skepticism of science. He trusts the Old Farmer's Almanac before the weather report. Weather is cyclical, he insisted, listing off a catalog of counter-arguments to climate science that I've heard around the country, including from many of the Dirty Dog mushers. The current warming can be attributed to volcanoes, they've told me. And sunspots. And solar winds. And the media doesn't report these things. None acknowledged that climate scientists account for these variables in their studies and readily accept the planet's natural climate fluctuations.

What the planet has not seen is as rapid a rise in temperatures, predicted to become warmer than they've been for millions of years, long before humans settled into their spaces and their sports.

"I still don't believe that man has been given the ability, no matter how proud they think of themselves, to completely control something that they're only on its surface for a very short time," Ron told me. "We're here, a blade-of-grass scenario," meaning that there might be seven billion of us, but we simply cannot have the impact that climate scientists are saying we have.

But while Mel politely agreed with her father that her recollection of colder winters in the past might be a trick of the mind, the truth is, data backs up her belief that Wisconsin winters are objectively milder than they used to be.

Starting in the 1980s, the frequency of winter freezes that plunged the thermometer below average was declining. By the mid-1990s, when Ron and his kids were running with the

Radio Flyer and Tiny's team, the cold autumnal spikes had nearly vanished. Ed Hopkins, Wisconsin's assistant state climatologist, told me that winters continue to be highly variable, with lots of snow some years and bare ground others, but he recently tallied up the length of the frost-free season since 1971 and found that in some parts of the state, it's increased by as much as three weeks.

And Now, a Word from ...

Dogsledding in Wisconsin has been changing for the past two decades for reasons aside from weather: the cost of dog food, the difficulty of finding long trail systems unimpeded by development or liability-averse landowners, the cost of fuel for the trucks to haul large teams. And one additional, significant change: the loss of sponsorship. It was sponsorship that had been keeping the races afloat, but sponsors started falling away as early as the late 1990s—often for weather-related reasons.

Ron spoke of one sponsor, the North Star Mohican Casino Resort over in Bowler, Wisconsin, that sponsored a dogsled race with a huge purse. "Then someone says ... we can make more money with a polka band," he said. "Bowler started to bring in live entertainment at the casino, instead of doing the race."

How infinitely appealing an indoor, climate-controlled event must be for a sponsor. The complete opposite of the iffy one-day mudfest that ended up being the Dirty Dog Derby this year.

"This is a weather sport," Ron said. "Your sponsors are expecting this much viewing of their product name, and if the weather is not conducive to that, you don't get the viewership."

"So if it's too cold, too windy, rainy," Mel continued, their conversation fluidly moving between them, "there are no spectators."

The prizes for the long-distance snow sled races can still be substantial—the Iditarod winner takes home $75,000—but as the purses have shrunk, sprint mushers are lucky if they win enough for gas money home.

The move from dogsledding to dryland racing to casino polkas is enough to make you wonder if we're doomed to become an indoor nation, seeking collective escape from an unpredictable world.

Many sports are suffering from the extremes in weather. Just as the sled dogs have their window where they can comfortably and safely compete, so do we two-legged athletes. It's difficult to play tennis when it's so hot your sneakers are melting on the court or you start hallucinating that you've seen Snoopy, as happened at the Australian Open a couple of years back. A study by the University of Waterloo's Interdisciplinary Center on Climate Change found that, unless carbon use plummets soon, a third of past Winter Olympics cities will be unable to host the event in the future because they won't get cold enough. Winter recreation sports are estimated to be a $12 billion industry in the country.

Loss of sponsorship and other factors are contributing to the decline in snow mushing in Wisconsin, but the greatest factor seems to be climate change. How long can the sport survive when the specter of uncertain weather is added to all the others? What is the fate of this sport—a healthy, life-affirming sport that people play instead of watch, that involves working in concert with animals instead of against them? Did the founders of the Iditarod think about the double meaning of their tag line, the "Last Great Race on Earth"?

Mel and Ron and I had talked enough. There were dogs outside, eager in waiting. Ron donned his coyote fur hat, its tail draped between his shoulder blades as the rest of it blended with the color and cut of his beard. Mel slipped on her Carhartt jacket and we headed out to the kennel where Keith had been setting out the lines to run the dogs with an ATV. It was too snowy for a cart and not snowy enough for a sled, so a motor would have to suffice. The dogs were rowdy with expectation, fervent to bound through the whitened forest, past Ron and Gail's garden, so recently put to bed, past the neat lines of the neighbor's fields, crisp in sepia tones.

"I love to watch them run, and run with them," Mel had said expectantly before we headed out. "To have them pull me and to be part of that team. And we're out there in nature, whether it's a beautiful sunny day, 20 below, raining, icing. We're appreciating what the planet has given us, and God's blessings that we're healthy enough to do this."

She was proving what her father had told me earlier about dog racing. "Nostalgia," Ron had said, "is a big part of this sport."

IN WEST TEXAS WHERE WIND POWER MEANS JOBS, CLIMATE TALK IS BESIDE THE POINT

Wind turbines bring jobs, tax dollars for new schools, income security for farmers and energy independence. To these Texans, climate change has little to do with it.

All along the straight-shot roads of Nolan County in West Texas, wind turbines soar over endless acres of farms, the landscape either heavy with cotton ready to harvest or flushed green with the start of winter wheat. The turbines rise from expanses of ranches, where black Angus beef cattle gaze placidly at the horizon. Here and there are abandoned farmhouses dating to the 1880s, when this land was first settled and water windmills were first erected. Occasionally a few pump jacks bob their metallic heads, vestiges of a once-booming oil industry still satiating an endless thirst.

Every industry creates an ecosystem around it. If the wind turbines that sprouted in West Texas were huge steel trees, spinning sleek carbon-fiber blades 100 feet in length, then the wind farms—including Roscoe Wind Project and Horse Hollow Wind Energy Center, some of the largest in the world—were their forest. Spread out across the expansive vista, invisible air currents feed the structures, their imperceptible roots extending out to the community that contains them.

"The wind industry has changed my life, and it's changed it for the better," James Beall told me as we stood at the base of one of those industrial versions of an old-growth tree, a 2 megawatt DeWind turbine that Texas State Technical College (TSTC) owns and uses to train its students to become wind energy technicians. While four of those students climbed 258 rungs of a vertical ladder to the nacelle, the streamlined mechanical housing at the top of the turbine, James and I awaited an elevator to take us up. From here on the ground, the nacelle, perched on its 300-foot column, looked tiny. It was actually the size of a bus.

As discussions around climate change in America have become partisan, so have those around kilowatts—but not here. There was money to be made, and these students—like the Texan Republicans in power over the last 20 years—weren't going to miss out on the chance to make it.

James is 39 years old, with a strawberry blonde beard and a voice smooth from his upbringing in the Nolan County seat of Sweetwater, also home to the West Texas campus of TSTC. As we waited for the elevator, we talked about the danger of working in the oil and gas fields—an industry that his father had made him promise, when he was 18, never to work in. James' father, a Vietnam vet, had worked in the oil fields when James was a boy, before finally getting disability pay for post-traumatic stress disorder, or PTSD, which dated back to his time in Vietnam when he'd been sent in to clean up the massacre of Hamburger Hill, James told me. But while Vietnam had broken James's father's mind, it was the oil fields that were the greater hazard for his body. It was there where he had broken his back and later his leg. It was the reason he had extracted the pledge from his son never to work in the industry—a tall order for a young man in Sweetwater, with few other opportunities.

Eventually James became an electrician—until 2003, when his son was born prematurely and needed someone to stay home with him. Since his wife made more money as a waitress at the local steakhouse, serving all the wind technicians who came in, it was James who became the stay-at-home parent. When his son was in school and he was ready to go back to work, he looked out at all those turbines popping up around his home and tried to get in on the boom, but employers wouldn't hire him because of his lack of experience. So he enrolled in the new wind technician program down the road at TSTC. Now, seven years later, he is not only a certified wind technician, but also a teacher in the program.

"The path I was heading down for a while" was not a savory one, James told me as the sound of the students climbing overhead reverberated around us. "I luckily escaped prison. Everybody has their vices at one point in time, and it's up to you to change your life and do what you need to do to be the better person. Wind helped me do that."

It was like finding Jesus. But with a steady paycheck.

The elevator arrived, and James and I squeezed inside the cramped metal mesh container. We rose as if we were water, inching our way up the interior of a great redwood tree by capillary action.

Wind's Ascent

Wind energy development in Nolan County dates back to 2001, when the first wind farms were constructed in the area. A perfect confluence of events led to the growth of the industry since then.

There was a supportive state government, led by Republicans George W. Bush and then Rick Perry, pushing for wind by putting the regulatory and infrastructure pieces in place to make it successful. The state's nearly autonomous electric grid meant no troublesome cross-border or federal approval was needed to get wind electricity from places like Sweetwater to the green-leaning urban markets clamoring for renewable energy. And then there were the Texans themselves, ever eager to use their land and diversify their revenue sources, especially as recurring droughts killed off the cotton and the livestock, and oil fields were either going dry or failing to pay for themselves. At the same time, federal incentives came (and went, and came again) in the form of production tax credits that helped the wind industry offset large investment costs.

If Texas were a nation, it would be the sixth-largest wind energy producer in the world. The bulk of that power is coming from the Nolan County region. And so the reddest parts of Texas are responsible for supplying upwards of 12 percent of the state's energy needs every month with clean, green kilowatts. Occasionally, as happened one day in the blustery month of October this year (a time when those energy-sucking A/C units are switched off and electricity usage is low), it provided more than half of the electricity to the state's power grid.

The Lure of Wind Industry Jobs

As the wind industry grew through the early 2000s, so did a desperate need for skilled labor. What emerged was the 2008 launch of TSTC's Wind Energy Technology program, where James enrolled in 2010 and where he returned to teach in 2013 after working in the field for a couple of years.

According to the Bureau of Labor Statistics, wind technician is currently the second-fastest growing job in America (beat out only by solar photovoltaic installer). By the end of last year, there were more than 100,000 jobs related to the wind industry nationwide, at least one-fifth of them in Texas. When the American Wind Energy Association (AWEA) launched a seal of approval for wind technician programs in 2011, TSTC was one of only three schools in the U.S. to receive it.

The students James teaches are a slice of the next generation of wind workers for an industry that, at least in this part of the country, has already established itself. They include veterans and women, those leaning politically right and left, environmentalists and climate change skeptics, the civically engaged and those who never vote. The clean energy component seems to be a bonus for some, but it was not the primary reason they chose this field. There is the laid-off gas worker who noticed all the wind turbines on the horizon and thought there must be an opportunity there. The English major who couldn't find a job and remembered how much she liked the outdoor work on her family's farm in the Texas panhandle. The two veterans who liked the element of risk and heights and the sweet spot of job independence and camaraderie.

Typical of the right-leaning students at TSTC was 31-year-old Scott Maxey. Scott had escaped his home in Redding, California, when he felt that the only job in town was working at Starbucks for minimum wage. He drove until his truck ran out of gas—which happened to be in Sweetwater. He was 25 years old, and he landed a job at one of the two big gypsum mills in town. Making sheetrock paid him more than double what he would have earned as a barista back home. But "at the gyp mill, the hazards there were just too much for my health," he told me. "I just wasn't doing well."

So in 2015, he found work in oil and gas, hauling the sand needed for natural gas fracking, which was on the rise in Texas alongside wind. But there was too much market variability with gas; within a year, Scott was laid off, and the guys he worked with told him to expect that to happen regularly.

"I'm looking around, and there's wind turbines everywhere," he said with a dusky laugh. "It just made sense. Why don't I work for the wind turbine industry?" Impatient to get back into the workforce, he opted for TSTC's one-year certificate program as opposed to the two-year associate's degree. When we spoke in November, he had one more semester left before he got his certificate, and he already had a job offer that would take him to Florida to work on solar, and then to California to work on wind.

In Scott's mind, clean energy has nothing to do with climate change, which he said has been with us "since the beginning of time. It happens naturally. I don't think human input is what's making it change."

And protecting the environment is not his motivation for getting involved in the industry. Sure, he said, it's important to be "responsible stewards" of the earth. "There's no reason to pollute rivers. There's no reason to go down and just mow down environments just because we can. Totally not okay." But he went into this field for one reason only: job security.

Scott is not alone in this—which is why the wind industry in West Texas is making for some very interesting bedfellows.

'Politics Falls to the Wayside'

Nearly three-quarters of Nolan County voters who showed up in November 2016 voted for Donald Trump, who has been more outspoken about reviving the dying American coal industry than about supporting the flourishing wind sector. But that figure only reflects who showed up at the polls. Of the four TSTC students I spoke to at the turbine tower, all of whom were investing their future in wind, only two had voted—a 50 percent rate that reflects the national average. There are always reasons not to vote. The Army vet was overseas, another student was busy. James was busy, too.

"I'm not really political," James said, "or when I do think about it, it's already over. I just get so lost with everything going on here, and home life. ... It's just overwhelming. Politics falls to the wayside."

While I was in Texas, one year after Trump's election, members of Congress were working on a tax bill that could affect the economics of wind energy in America. The fact that crafting the tax bill was in many ways an exercise in horse-trading behind closed doors seemed to both support and refute the reasoning of everyone who had decided to stay away from the voting booth.

But what happens in D.C. comes to roost on the tips of the turbine blades of Sweetwater. Just as the students of TSTC were seeking certainty for their futures, so too has the wind industry sought stability in the federal financial incentives that help offset initial capital costs. After years of fluctuating tax credits, a bipartisan effort by Congress in 2015 locked in a five-year Renewable Electricity Production Tax Credit (PTC) for wind development. It provided a 2.3-cent tax credit for every kilowatt of electricity produced, with a gradual decline over a five-year period. By mid-November the House version of the tax bill aimed to eliminate the credit immediately. It would have meant a complete mid-stream change in the financial game plan for wind energy companies and the financiers who back them. What American businesses, towns, and citizens want is the opposite of these oscillations; they want certainty.

Business interests in Sweetwater were watching these changes on Capitol Hill with some trepidation. "If it affects the industry, [even] if it doesn't affect us today, it will affect us at some point," said Ken Becker, executive director of the Sweetwater Economic Development Corp (SEED). We were sitting in his downtown office, on the second floor of an historic Spanish Colonial Revival style home that serves as the Chamber of Commerce, as he told me how wind energy had bolstered the local economy.

"In pre-wind, our county taxable value was $500 million," Ken explained. "In 2008, it was $2.8 billion," a five-fold increase that translated to new schools and grand expansions at the local hospital. That's money for the town, but also a steady income for local landowners, some of whom earn up to $1,000 per month from having a single commercial turbine on their property—and most of the region's world-class wind farms are dotted across private land. Many say they're "not sure they'd even have the ranch today if the wind didn't come on," Ken told me.

The wind construction boom has slowed—or, perhaps more accurately, morphed—as the industry enters its second generation. Sweetwater has a population of about 10,000, and back at the peak of wind development in 2008, as the economy was crumbling, it was estimated that 18 percent of the town's working population was employed on wind projects, Ken said. Now, he acknowledged, the number of full-time jobs in wind technology is a small fraction of that, but only if you disregard the hundreds of peripheral jobs that rely on the wind energy sector.

These complementary industries are the ecosystem that wind power belongs to—and its reach is growing. Repowering, which vastly increases efficiency by either replacing old turbines for more powerful ones or upgrading components, means more megawatts with the same footprint. It also means a whole new category of jobs. While I was there, evidence of these peripheral industries was everywhere. I watched 80-foot blades swapped out for ones twice as long. (The production tax credits helped these efforts, too.) I visited Global Fiberglass Solutions of Texas, which was setting up shop in an old aluminum recycling plant to process the decommissioned blades—which were being amassed in a 10-acre field—into building panels and other materials.

Across the street from Global Fiberglass Solutions was the Argentinian company EMA, which builds electrical breakers specially designed for renewable power systems and employs about a hundred people.

And down the road in Roscoe, in an old mercantile brick building inscribed with "Shelansky's Dry Goods," beside a high curb designed for horse carriages, teens at Edu-Drone were learning skills that would earn them FAA licensing as drone pilots. They were simultaneously earning a high school diploma and an associate's degree, aiming for a career that could have them use the flying cameras to inspect wind turbine blades for damage from hail or lightning strikes.

The Wind Will Always Blow Here

Back at the TSTC turbine, high above the mesquite below, the elevator shuddered to a stop at the top. Well, almost at the top. While James, who had a hurt shoulder, waited below, I clipped in my lanyard line and climbed up the last 40-foot stretch to get up to the nacelle.

The four students who had climbed up by ladder were already there, along with Billie Jones Hudson, another TSTC wind instructor. Billie, 42, whose long brown hair was pulled back in a tight ponytail and tucked down her shirt, had been a prison guard and an elementary school art teacher before working in wind. Together we stood amid the hydraulics, cooling systems and electrical boxes of the nacelle. It was all basic mechanics, but supersized. Big bolts, smeared with grease, attaching one massive piece of metal to the next. There was the slightest of sway to the entire structure, which one student, Kaitlin Sullivan, 25, equated to a boat.

With one more hoist, I could look out the hatch that opened up to the sky, the blades reaching to infinity, airplane wings on end. This is what everyone I spoke to in the wind industry loves the most about this work. It is the exact opposite of descending deep into the earth to mine coal in shafts no taller than the height of a small child. There is risk to wind work—an arc flash, a long fall—but never the prospect of being trapped underground for days as there is with coal. There are no oil leaks that can't be staunched. It felt like freedom. Energy freedom. Professional freedom. Economic freedom. Our turbine's blades were still, but I could see dozens of others spinning along the horizon below, the wind that blew through them altered, molecules shifted.

"It's breathless," was how student Johnathon McCarthy, a 28-year-old West Virginian who served in Afghanistan as a Marine, described being up there. "You see other turbines around you spinning. You see the cars that are just this small. You see the people down there and they're even smaller, like ants. You just feel like you're on top of the world."

The best places for wind are often the places that are struggling to keep rural communities alive.

What was happening in Nolan County proved that the debate about how we generate our kilowatts doesn't have to be about climate change. It could be about embracing whatever clean energy options are available to help make small-town America economically viable. In this deeply red place, it was the embodiment of President Barack Obama's all-of-the-above strategy. At the close of 2016, 86 percent of the country's onshore wind turbines were located in Republican districts, according to the 2016 U.S. Wind Industry Annual Market Report. Indeed, Republican Sens. Rob Portman of Ohio and John Thune of South Dakota were some of the primary advocates responsible for keeping the PTC in place in the final version of the tax overhaul bill, which was signed Dec. 22.

Too soon, it was time to leave our aerie. Kaitlin sung the words to "When It Rains" in a clear soprano voice, the confined space of the nacelle a perfect sound chamber, as we all worked our way down. I climbed back to where James awaited me, and we returned to our constrained positions in the elevator.

"What was advanced yesterday is no longer advanced today," he said as we slowly descended, "and what's advanced today will no longer be advanced tomorrow. Same way with wind turbines; they're always advancing." Periodically, light slipped through into our confined space, illuminating his face.

"This is a totally different world. I believe this is the future," he said, pausing for a moment. Then he added, almost to himself, "I hope it's the future."

FLY-FISHING ON MONTANA'S BIG HOLE RIVER, SIGNS OF CLIMATE CHANGE ARE ALL AROUND

Nature is a powerful economic driver here, and livelihoods depend on cold water and healthy fish. People know it's warming, but few will say that's climate change.

Anyone who takes fly-fishing seriously behaves like a scientist. These anglers are biologists, knowledgeable in what's eating what, when and how. They are hydrologists, studying riffles and stream flow. They are naturalists, observing clouds and sunlight and the circulation of air as their rods flick back and forth across the big sky. They are, in a sense, climate scientists. And some, but not all, are deeply concerned about the effects of a warming climate on the cold-water species that inhabit blue-ribbon trout streams.

But to the extent that they act as climate scientists, partisan politics plays a role in many anglers' understanding of climate change. Here in Montana, with pristine rivers that are home to some of the best fly-fishing in the country, a majority of votes went for President Trump—and climate change is considered by many of them to be a natural phenomenon beyond human control. Nonetheless, climate change is having a profound influence on fly-fishing, from the timing of insect hatches to the long-term survival of the fish that give this sport its meaning.

The classic account of angling in Montana depicted in Norman Maclean's *A River Runs Through It* implanted iconic images in the collective consciousness, and they are not false. But will they survive the century?

On an early May day, on the upper reaches of the Big Hole River in southwest Montana, fly-fisherman Craig Fellin is in that quiet contemplative state of the experimental scientist as he steps out of his Suburban. Already he is studying the swirl of deep eddies on Grayling Pool, searching for the movement of insects. Before he casts his luck into the river, he shows me how he decides which fly to attach to his rod. First step, he says, is to put "your nose on the water."

Craig, 71, has a neat swoop of mustache and a calm, deliberate air. He founded the Big Hole Lodge nearly 35 years ago, putting to use a degree in philosophy and a lifetime of fishing, begun during his childhood on family trips to Canada to fish for walleye and pike. He is a Vietnam veteran, and a lifelong Republican. He is also convinced that climate change is affecting the pastime and livelihood he loves, from the trout in his backyard to the steelhead he seeks in the Pacific Ocean. What he can't figure out is where the outspoken conservationists among his fellow conservatives have gone.

Seasons of Deception

Like preparing for a sacrament, or a science experiment, Craig slips on waders, assembles his 9-foot graphite Orvis rod, and slips on a vest laden with tools at the ready. Even a nose on the water doesn't reveal much today—there are few insects active—so he mulls over a fly box that opens like a book, revealing 32 compartments with tiny transparent spring-loaded doors. He contemplates an array of flies—hooks and feathers spun together with thread and wizardry—and settles on a Parachute Adams dry fly, its light and dark body made to mimic an adult mayfly. I watch him as he secures it to his line, cinching off the knot by tugging the line between teeth and fingers.

Stepping to the edge of the river, Craig flips his rod to get just the right momentum on the light line, landing the fly upon the surface and then lifting it up again. He works a spot and then continues upstream. It is a moving meditation. A continuous motion underlain by deep stillness. Not unlike the stillness in the depths of Grayling Pool, where the trout seem to be laying low, ignoring the rush of the current and the temptations offered overhead. After a couple hours, Craig still has not found his honey hole, that dream spot where the fish are abundant and biting. Today he will not hear the zip of a line when there's action on the other end, or hold a slippery creature for a moment before releasing it.

"Catch and release" is the common practice here. The sport of fly-fishing is not about securing food, but about the thrill of the chase, the skill of the catch, and about communing for a time in this "Last Best Place," as Montanans call home, with snow-capped mountains dropping to lodgepole pine forests and opening to grassland valleys erupting, just now, with wildflowers. The trout, Craig explains, giving up after a couple of hours, mainly bite when the water temperature is between 45 and 65 or so. They are, in a sense, piscatorial Goldilocks.

Today the water is too cold, he suspects. It was "an old-fashioned winter," Craig says, with record snows. But nowadays the old normal is an anomaly: For the past 30 years, the state has been warming—and at an unusually fast clip. Despite its long cold winters, Montana's average temperature in 2016 was 3.5 degrees Fahrenheit above its 20th century average. That's double the warming of the planetary average from the same year. Since 1987, when Craig first opened the Big Hole Lodge, only three years have been cooler than the average temperatures of the last century. The chart of this warming progress is a jagged sawtooth—cold years and warm ones—but reaching ever upward.

Even this year's hefty snowpack, which feeds the river much more than the occasional summer rain, no longer assures a good summer of fishing. "It was actually up in the low 70s last week, and we lost 25 percent of our snowpack in one week," Craig explains. "Twenty-five percent in one week," he repeats. "Unbelievable. For April, that's very unusual."

The ideal scenario is good snowpack followed by a gradual descent into summer, so the meltwater is meted out steadily. This goes for anglers, as well as for Montana's ranchers and farmers, who fear the droughts associated with climate change, and its firefighters, on the alert for the wildfires that feed on dry conditions.

"A long, slow release of mountain water is always preferred, but isn't always delivered," the water supply outlook report says. When winter leapfrogs spring straight into summer, the water melts fast and furious and then is gone, leaving the second half of summer parched. While I was fishing with Craig in southwestern Montana, the Milk River in the north was flooding due to rapid melting, causing the governor to declare a state of emergency. At a weather station near the Big Hole River, nearly 90 percent of the snowpack disappeared in April. So much snow, gone too quickly. And when that cold water is gone, rivers flow low and warm up fast. That is a disaster for cold-water fish.

The Big Hole River is feeling the effects most dramatically. Locals call it the "Last Best River," undammed and wild and gorgeous. It is trout heaven—rainbows with their iridescence, browns covered in spots, and the native westslope cutthroat with red slashes along their necks. The river also has mountain whitefish and the very last of the Lower 48's native stream-dwelling fluvial Arctic grayling, sleek and silver with a blue-spotted dorsal fin that flows like a sail. The thousands of fish that ply each mile of river feast upon a succession of stonefly hatches, some as small as dust motes, others the size of your finger. Both predator and prey are dependent upon cold, clear waters for their survival.

Conservare: 'To Keep, Preserve, Keep Intact, Guard'

Craig is not only a conservative, but also—and maybe even primarily—a conservationist. He speaks of the Republican presidents who signed the Clean Air Act in 1970 (Richard Nixon) and its amendments in 1990 (George H.W. Bush). Although Nixon initially vetoed the Clean Water Act in 1972, overwhelming bipartisan support in Congress overruled him. He also established the Environmental Protection Agency. All these efforts helped address pollution in Montana, and all across America.

Where have those voices gone? Craig asks. He doesn't hear them on Fox News. "Nobody talked about conservation and the environment" during the last election, says this self-declared "frustrated Republican." But even his deep concern for the land and waters he loves isn't enough to sway his vote, which he bases on more than the single issue of climate action.

"I voted for Trump," he says, "for the Republican ideas of smaller government and less taxes and more pro-business."

Yet despite his party's refusal to embrace efforts to reverse climate change, Craig has kept an open mind. He watches Fox, but also seeks out information from many sources, such as a recent episode of 60 Minutes, where he learned about ocean acidification, another aspect of climate change that impacts fisheries.

But an event last year was the clincher for Craig. A friend invited him to a talk by geologist George Brimhall at The National Exchange Club, a gathering of businessmen in Butte. George gave a Powerpoint presentation, and somewhere between talking about Humbug Spires and Butte ore deposits, he focused on climate change. Showing temperature graphs drawn from data reaching back 500 million years, he explained how the climate has always been changing, with distinct warm and cold periods in the past. Natural climate cycles. That made sense to Craig.

But now, George added, we're supposed to be on a cooling trajectory, not a warming one, a *rapidly* warming one. "We should have been heading back into our next ice age," George said, but because of our use of fossil fuels, "we stopped nature from doing that." That was the point that lodged in Craig's mind, troubling him still. "He showed me that it's for real," Craig says, recalling the talk.

Craig's reaction to this data was unusual among the people I met along the Big Hole. Others recognize that change is happening—the effects are hard to deny—but these observant anglers come to more predictably modern-day Republican Party-platform conclusions about its origins.

"There's always changes, everywhere," says one of Craig's neighbors, Frank Stanchfield, who founded his own fishing outpost, Troutfitters, on the Big Hole River around the same time Craig opened his lodge. Montana is warmer than it used to be, he tells me: "We used to see 50 below several days every year, and we rarely see 50 below ever, anymore." But he says a human lifetime is but a blink of the bigger picture. As we sit together on the banks of the Big Hole, Frank clutches an unlit cigar in his hand, placed loosely over his stomach. "I don't think it's manmade," he says. "I don't think there's anything we can do to change it ourselves."

On another day, another fly-fisherman tells me something similar. "It's changed, there's no question about it," says Jim Hagenbarth, who is also a rancher and, along with Craig, a founding member of the Big Hole Watershed Committee, which has been working to improve river conditions since 1995. "It's Mother Nature," Jim tells me. "The climate has always changed. I think it'll swing back the other way."

Even Mark Thompson, the president of the George Grant Chapter of Trout Unlimited in Butte, says climate change isn't on their agenda. "Our sole purpose for Trout Unlimited is to conserve and protect cold waters fisheries," he tells me as we cruise in a pickup truck along MT-43, which winds along the Big Hole River. "A lot of the species, particularly here in Montana, don't do well with warm water," he acknowledges, but as for climate change, "it's not something that's ever come up. We don't talk about global warming in our chapter meetings." Instead they focus on the immediate: creek restoration projects and what can be done on the ground, now. When I ask him if he thinks there's a role for the federal government to play in reducing carbon emissions, he says, "No," before I can finish the question.

No matter what those on the banks of the river perceive, the reality of the fish in the water is another thing altogether.

Changes in the Water

One way to measure the possible impact climate change is having on the Big Hole River is to monitor the health of its fish. That's the job of Jim Olsen, a fisheries biologist for Montana Fish Wildlife & Parks. He tells me that disease outbreaks related to warm waters are on the rise, including a fungus called saprolegnia and a parasite that causes proliferative kidney disease (PKD).

Late one afternoon, I catch up with Jim along the watery willows of Bear Creek, a tributary to the Big Hole River, where he is collecting brook trout in a bucket to test in the lab for disease. "We're seeing changes for sure," he tells me when I ask him about climate change. "Our spring seems to come a little earlier and fall seems to last a little longer."

Four years ago, two warm weeks in October led to "pretty

significant die-offs" of Big Hole trout, Jim says. The brown trout were spawning, which put a stress on their bodies that, coupled with the heat, made them unable to fight off the saprolegnia fungus, as they could in cooler water. The population took a dive.

Pinning the deaths "exactly on warming temperatures is ... " Jim starts to say, then pauses, the ever-cautious scientist. He starts again. "I don't know if we can do that yet. But for certain we're seeing changes in the fishery in the last five to 10 years." He explains that the species of fish predominating along the river have shifted dramatically in that time.

"Ten years ago, a brown trout in the upper end of the Big Hole was rare; it was all brook trout," Jim says. "Now we've seen that almost completely flip flop." Brown trout have been in the Big Hole for almost a century, but it's only now that they've come to dominate. He's unsurprised by this, given that brown trout can tolerate warmer water than the brookies. "I don't have any other explanation other than temperature."

If an angler just wants to catch a fish, any fish, then these shifts may not matter much. Many of the fish associated with Montana rivers were only introduced in the last century, when stocking rivers was as common as slipping cans of Campbell's soup on grocery shelves. But people have their favorites. Craig likes the challenge of brown trout; they're reclusive and hard to catch. Many are fighting for the cutthroat trout, a native trout that Lewis and Clark feasted on when they passed through in 1805. Today, cutthroat are down to 6 percent of their historic range.

And then there's the fluvial Arctic grayling, which was once found across Michigan and Montana. Now, the Big Hole River is their only habitat in the continental United States, with only about 200 breeding pairs. As a cold-water species, they remain abundant in Canada and Alaska, but in Montana they can only retreat so far to higher cooler elevations if temperature is a pressure on them. Like many species across the planet, there will come a time when there's nowhere left to go.

Luckily, the Big Hole so far has been spared the massive fishkill suffered on the Yellowstone River in 2016, when PKD, exacerbated by warm water, killed thousands of fish and shut down 100 miles to all recreation. But multiple sections of the Big Hole River frequently have restrictions or outright closures due to low water flow and warm water during peak fishing season. In the last four years, the upper stretch of the Big Hole had hundreds of days when these limitations were in effect.

That means less fishing, and more anglers crammed into the river sections that remain open.

Montana Fish Wildlife & Parks Region 3, which includes the Big Hole River, is only 12 percent of the state's land area, but it provides more than a quarter of the state's angling opportunities, and that translates to dollars. Tourism is a substantial part of Montana's economy, and many people who come, come to fish. Two-thirds of Montana's wildlife management budget comes from fishing and hunting revenue, so anything that makes either activity difficult (or impossible) has an economic impact. The angling industry alone generates around $300 million each year.

Resilience, and When It's Not Enough

Regardless of their politics, those who fly-fish have been acting like conservationists for more than half a century. Decades ago, when fish stocks became thin, the practice of "catch and release" was embraced. In the 1970s, fisheries management switched to supporting habitat instead of stocking. When severe drought hit in the late 1980s, the governor tasked ranchers and outfitters to come up with a water management plan and, eager to keep the federal government out of their affairs, they did.

These efforts have paid off. Cattle are kept off riverbanks to prevent erosion. Willows were planted to stabilize banks, shade the water and create cool hiding spots. And when the water gets too warm, causing fish stress, anglers stay away. If water temperatures hit 73 degrees Fahrenheit over three consecutive days, voluntary "hoot-owl" restrictions go into effect, limiting fishing after 2 p.m., when the day is at its warmest and the fish can't handle the stress of being caught—imagine going for a run on a hot day. The best of anglers will spend time swishing the fish they catch back and forth in the water to oxygenate its gills before releasing it.

But will habitat restoration and hoot-owl limitations be enough? Jim Olsen, the fisheries biologist, is not sure. "If climate continues to change and get warmer," he says, "there may not be anything we can do in these lower reaches which are warmer and have lower flows." While currently, warming seems merely to be shifting species around and causing the occasional fishkill, the long-term future looks grim for Montana's trout population. Studies show that nearly half of all trout habitat in the interior West could be gone in the next 60 years. Brook trout could lose more than three-quarters of their current range. Migratory bull trout could be almost entirely wiped out. By then, the Arctic grayling will have vanished completely from the rivers of the continental United States.

The Seen and Unseen Nature of Climate Change

In *A River Runs Through It*, Norman Maclean writes, "All there is to thinking ... is seeing something noticeable which makes you see something you weren't noticing which makes you see something that isn't even visible." How do we connect the links between the seen and the unseen? How do we wrap our heads around something as complex as climate change? Something so immense it can seem unbelievable?

In a way, that's what a fly-fisherman like Craig is doing all the time. As he and I chat over a picnic lunch on the banks of the Big Hole, he continually scans his surroundings. Finishing his ham sandwich, Craig spots something, and reaches out to grab it.

"Whoa, whoa, here's a stonefly," he says, holding it out in his palm. "Isn't she beautiful?" He identifies it as a golden stonefly, carefully pinching off an ant latched to the stonefly's hindquarters. He then comments that its appearance is a month early. Around us, the sun is shining brightly. The snow is melting quickly. "That could be a precursor," he continues, troubled, looking at the creature before letting it go.

"Even though it was a cold winter, this is a sign of climate change—that things are warming up sooner," he says. What it means for the future of fly-fishing, and the cold-water species that anglers get such joy pursuing, remains to be seen. Or unseen. As those with rod in hand decide, on the river and at the ballot box.

THE FLASH DROUGHT BROUGHT MISERY, BUT DID IT CHANGE MINDS ON CLIMATE CHANGE?

Ranchers in Divide County, North Dakota, rely on the rain. Last year the rains failed, and the temperature shot up. 'The crops just didn't come out of the ground.'

DIVIDE COUNTY, North Dakota — I walk in the front door of Byron Carter's house as others are entering in the back, and Koda the dog can't decide which way to direct her barking. I'm in Divide County, North Dakota, but borders seem a little meaningless here. Last summer's drought, which was calamitous for Byron and the other farmers and ranchers now filing into his kitchen, leaked over into Canada, Divide's border to the north, and Montana, to the west. By April of this year, they're on the cusp of a new season, and Byron has gathered his neighbors—defined as anyone living within a 30-mile radius in this sparsely populated corner of the state—so we can talk about drought and climate change.

Drought is an especially wily adversary. As an officer of the North Dakota Department of Emergency Services told me recently, "You can't put up a sandbag wall to stop a drought."

In Divide County, agricultural producers are especially vulnerable to the effects of drought, since they depend on dryland methods. Dryland farmers use no irrigation. Instead, they rely wholly on rain: to initiate the lush growth of little bluestem and other pastureland grasses that will sustain their herds through the summer, and to secure the hay harvest that will get the herd through the winter. Not to mention the rain they need for their wheat, barley and pea cash crops.

In 2017, ranchers were optimistic when they put their cattle out to graze in late spring. There'd been record snowfall over the winter, and regional forecasts weren't calling for any drought conditions in their northwest region of the Great Plains. By May, though, concerns were rising. Rain failed to come, and the good winter moisture evaporated into a cloudless sky. By July, two-thirds of the pastureland in the Dakotas was in poor condition, and across the High Plains, from Kansas up to Canada, temperatures were above normal while precipitation was low—perfect conditions for what's known as a "flash drought," sudden and severe.

By the first of August, the USDA reported that nearly three-quarters of North Dakota's topsoil was desperately bereft of moisture. Part of Divide County was at the most severe drought level, and 60 percent of the state was facing some level of drought. It was the state's fourth-driest summer since record-keeping started in 1895. Ranchers hauled water to their herds and vied for hay donations that flowed in from other regions after the state opened a hay lottery. Anything to supplement the feed of the hungry cattle.

What happened? How had it happened so fast? And would it happen again?

"Probably some of the worst droughts I've ever had," says Greg Lee, a soft-spoken and gentle-eyed rancher who's been working the land here for 40 years. He remembers the staggering drought of the late 1980s, but he never experienced a year like 2017. The land "looked like this," Greg says, tapping his hand on the table's hard surface. "The crops just didn't come out of the ground."

"We had six-tenths of rain about the 16th of June," Byron remembers about last year's rainfall, hopping up on the kitchen counter. The rest of us sit at his kitchen table, drinking coffee as our boots lay piled in the corner.

"You're measuring in fractions of an inch?" I ask.

"Yes," he says, adding that it's easy to remember last year's weather events, which every farmer-rancher keeps track of religiously, because "there weren't any." He recorded a total of 2 inches of precipitation all year compared to the average of about 12.

"Every day got drier," Byron says, as the men around the table nod in agreement. "It's very unusual. Very unusual!"

You can feel the trepidation in the room. The prairie potholes of this northern Great Plains region are filled with fresh snowmelt, glistening in the late April sun, but it's a mirage of sorts. When the frost line melts, the "land takes the drink" and the potholes mostly drain, but not enough to replenish the soil.

While this region is one of the coldest parts of the continental United States, North Dakota is experiencing one of the greatest temperature increases in the Lower 48. An incremental uptick across the region means that it's now running about 3 degrees Fahrenheit over last century's average, a treacherous trend that can dry the land out too quickly in a region where ranchers rely on the rain to keep their rolling hills greens and their herds healthy.

Land of Extremes

North Dakota is a land of extremes. On the heels of last year's drought, the region experienced a long, cold winter. All the ranchers in Byron's kitchen recall the misery of tending pregnant heifers in early 2018 at sub-zero temperatures in the middle of the night. (One of them, Jim Reistad, recounts an episode of helping a cow give birth, during which he ended up with a shot of fresh manure down his shirt. "I wonder if these frickin' New Yorkers know just what we do to put that steak on the table," he says, with a wry smile.)

The variability of the weather is a lively topic in Byron's kitchen, as his rancher friends speak of these highs and lows. The men tell me about the good wet decade prior to last year's dry year, and about how Divide County was under water in 2011, rainwater submerging roads as it spilled over from the natural prairie potholes. "We had production that year," Byron says. "What crop we got in was good. But in a drought, no production is no production." The region, he says, seems to go "from one extreme to the other."

"Every year is different," someone says, and there's assent all around.

"I cannot get on board with global warming," another rancher says. "I do believe we have cyclical stuff. We are on a little bit of a warm-up. But we're probably going to go the other way, too. It's been doing that for years, for eons."

Things Change

Earlier that morning, I'd met with Keith Brown, who was the agricultural extension agent for Divide County for 31 years before retiring a couple of years ago. Beneath the brim of a camouflage hat, he spoke with a quiet equanimity.

"This whole area was covered by ice at one time," he said. "This area was covered with swamps at one time; that's where the coal and the oil came from." Indeed, this region is part of the Bakken Formation, which boomed and then busted with oil and gas production, though rigs and gas flares are still common, along with elevated land prices. "Things change," Keith said.

"I've seen the data that indicates, yes, we are warming, but there doesn't seem to be a lot of agreement," he added. "Some say that it's the result of man's activity and others say it's just a natural occurrence."

I've heard this sentiment across Divide County and across our divided country. The confusion is not surprising, given concerted efforts to make climate science cloudy to the public. What is perhaps surprising is to hear the "some-say" argument from cooperative extension agents like Keith, who are trained in agricultural science to understand how climate affects farming and ranching productivity and to be the experts within their communities. His statement suggested he was unaware that 97 percent of climate scientists agree that human-caused climate change is happening. The American Association for the Advancement of Science and most other prominent scientific organizations have officially stated their agreement with this consensus.

To understand what's been happening to the atmosphere since the start of the Industrial Revolution, scientists use many of the same data sources that farmers and ranchers turn to first thing each morning to see what the weather is doing. They're just assessing it with a longer time frame, and adding deeper layers of paleoclimate information for greater understanding.

But in this place of extremes, the impact of climate change can be difficult to track. As the North Dakota state climatologist Adnan Akyuz told me, "It's hard to pinpoint the attribution," since many factors contribute to extreme weather events. But he recognizes that climate change is having an impact and didn't hesitate to say that the state is "at the epicenter of these temperature increases." Like Keith, Adnan referred to North Dakota cycles of wet and dry, harkening back to the ever-present memory of the 1930s Dust Bowl.

"I am not ignoring the impact of the increasing temperatures and the changing climate," he said. "It's a big factor, because increasing temperature means air becomes drier and thirstier for moisture, and that moisture has to come from the soil." He also explained how a warming climate exacerbates floods as much as droughts. "You expose yourself to more frequency of floods," since warm air holds more moisture, so when snow or rain falls, it can come as a bombardment. These familiar extremes getting even more extreme fit climate models that project more frequent and intense droughts coupled with more severe rain storms, according to the National Climate Assessment.

Especially since the droughts of the 1980s, farmers themselves have responded to the vagaries of weather by changing some of their practices. The farmer-ranchers I met consider themselves true environmental stewards. They know the land intimately and point out the changes they've made in the last generation.

"We're much more efficient than they were 40 years ago," Jim says. "We kinda feel like we are doing our part."

Many have adopted no-till farming, which means planting directly in the stubble of the last year's harvest instead of turning over the soil with a plow. This approach helps leave the soil structure intact, allowing it to hold more moisture—and more carbon. Soil is the unsung hero when it comes to the planet's carbon storage, serving as a carbon sink greater than the atmosphere and all plant life combined. Only the oceans contain more carbon.

As Byron puts it, "It's pretty tough to find anybody besides a farmer-rancher that loves Mother Earth as much as we do, because we work with it every day."

Reserves and Options

But these men walk a thin line.

"We're a month away from total liquidation here for a lot of guys," Jim says. If it doesn't rain and "green up," he says, there will simply be nothing for the cattle to eat.

The only real luck of last year was that the soil held moisture reserves from prior good years. By April, those soggy savings are spent, and the 18-month reserves of hay that ranchers like to keep at the ready are depleted, especially with the hard winter.

"The reserves are gone in this region," Byron says. "There's a hay shortage. There's a straw shortage. Everybody always has a little reserve, but that's getting exhausted."

"A lot of cows are hitting the road," someone says of the need to sell off some of their breeding herd, which is the rancher's capital asset. "We have no choice at that point."

A couple of days after my visit to Byron's, I see this fragility in action at the Sitting Bull Cattle Auction down in Williston, the largest nearby city. Most of the year, buyers and sellers arrive weekly from around northwest North Dakota and eastern Montana to buy and sell cattle.

Haugland's Action Auction is running the show today. Haugland's is a family business, with the father-daughter team of Butch and Amber taking turns as caller and spotter. Amber, in her mid-30s, is not only an auctioneer but also a rancher, lodge owner, and an insurance agent with Farmers Union Insurance. She has a great big smile and a head of tight dirty blonde curls and impeccable makeup. For fun, she wrestles greased pigs. (And wins.)

As an insurance agent, she handled a lot of paperwork last year as she quietly filed her neighbors' claims for crop losses. But as part of the auctioneer team, she gets to hoot and holler.

"Yah!" Amber bellows when she spots a nearly indecipherable gesture from bidder number 421. During her turn as auctioneer spotter, she constantly scans the afternoon crowd of about 45 men, women and children filling the steep bleachers as two men usher groups of Red Angus cows across the dirt-floored stage. The bidding ramps up in a tongue-twisting tangle of numbers before the cattle are sold to the highest bidder. The spotter hollers when the bidding gets hot. But it's not as hot this year. More cattle for sale, but more buyers sitting on their hands.

"The folks who maybe would have picked up an extra bull or two to have for their breeding season this year just aren't as willing to put that money down," Amber tells me when the auction wraps up. "They're kind of scaling back and just tightening up for what could be another difficult year."

Amber and her ranching neighbors have a host of concerns about what happens when drought hits. They have to test the prairie pothole sloughs that serve as natural watering troughs, to make sure low warm water hasn't flushed green with toxic algae. They have to monitor the hay harvest's nitrate level, which elevates when the plants are stressed. They have to manage ever-changing tariffs (and threats of tariffs) that affect everything from the price of steel in their farm machinery to the price of grain. And they have to fret about the fate of the Farm Bill, set to expire in October, which now subsidizes the crop insurance that farmers and ranchers depend on every year. As much as I heard about frustration with excessive government regulation—and I heard it a lot—even Amber, a conservative, admits that "without government intervention," the economics of farming "would collapse."

While the federal supports kick in when droughts hit, government interventions on the local level also help keep farming and ranching viable during dry times. The state, for example, has spent hundreds of millions to secure long-term water security by digging wells and running pipelines to pastures, while enacting immediate drought-relief measures such as hay lotteries. The need for all this support could eventually be lessened if climate change were addressed.

But the cattle industry in the state has been fighting government action on climate change at every level. Leave it to us, the North Dakota Stockmen's Association (NDSA) said in a resolution approved in 2015. "WHEREAS, cattle producers are ardent stewards of all natural resources including land, livestock, water and air," the resolution began, "the NDSA works to reduce or eliminate efforts to regulate greenhouse gas emissions under the Clean Air Act, related climate change legislation or promulgated rules."

Amber also tells me that she's not concerned about climate change. Like Byron and his friends and so many other North Dakotans I spoke to, Amber refers to the familiar extremes when talking about the climate. She remembers when the weather was so wet in 2011 she had to truck her cows out a back way to get around flooded fields. And she remembers being a kid in the drought-stricken '80s, when "you would run and grasshoppers would be flying" because it was so dry. In the face of extremes like this, she told me, you hunker down and manage risks.

And "you pray for rain," she adds.

"You have to at least acknowledge and address that there are changes or things going on," she says. "We had a year last year that was really, really dry. Our hay crop was like a third of what it typically has been. It was devastating to this area." But, she says, "there's a deep distrust in terms of what you read" about climate change, and she questions scientists' motives.

What she does trust is her own experience. "We are out there every single day, constantly with those animals, so if anybody is an expert on what's going on with the weather, it's probably the people connected to it and agriculture."

Connection in Divide County

"Belief" in climate change falls mostly along party lines in Divide County. During my time there, I met mostly Republicans; the county is clearly a red one, with 71 percent voting for Trump in 2016. But I also met a few Democrats, as well as Independents like Sydney Caraballo, who says she has voted for people on both sides of the aisle.

Sydney, in her early 40s, was born and grew up in Divide County. Her great-grandmother homesteaded here back in 1905 after making a land payment of $14. Sydney spent 20 years away, and after marrying her husband Kevin and having three children, they decided to return to her home ground to raise their family. They returned in 2013, taking over Glasoe Angus from her parents and carrying on the tradition of raising registered Black Angus cattle for breeding. They also farm wheat, barley, lentils and more on thousands of acres. They haven't looked back.

Sydney believes climate change is "absolutely" human influenced, but she is concerned about what she calls "fear mongering on both sides" – especially on social media. "Compromise and solutions become lost in the rants."

Make our home places better, she argues, by which she means both our ranches and our planet. "In terms of climate change, we need to adapt, and continue to adapt, and try to do things that make this a survivable place for the wildlife, for us, and for the livestock," Sydney says. "It would be foolish for me to engage in practices that would put me out of business. We're a family farm. This land isn't just for me, it's for future generations." She adds: "We're trying to feed the world."

On a Friday evening in April, the sun is setting as I stand with Sydney during a quiet moment. She and her family have just brought the cows in from the field. Behind us is a pile of hay much smaller than Sydney would like. And in front of us, the work done, her daughter Riley Jo, 8, is cow whispering.

The child is standing, in this field, on this land, in her sparkly shoes and floral tights, still as a heron, with her hand outstretched. She is willing three Black Angus calves and their wary mothers to succumb to their curiosity, and they do, sniffing and stepping forward, retreating, returning. She takes steps closer to them, moving slowly, with a hyper focus not often observed in eight-year-olds. And in the gesture, I can't help but think about North Dakotans and their intimate relationship to the land and the food they help it produce. In that moment, I see Riley Jo and the calves as proxies in our bitterly partisan debate about climate change in this country. Could a fragile connection, fleeting and tenuous, be made unbreakable with time and more trust—in the science that shapes the world we inhabit and the stories we tell each other, whether online or around our kitchen tables?

THEY KNOW SEAS ARE RISING, BUT THEY'RE NOT ABANDONING THEIR BELOVED CAPE COD

Lifelong residents are building higher with each flood. But while they contend with climate change, some say they aren't sure what to believe about the cause.

"It flooded in early January, and then it happened again two or three months later," says Matt Teague of Barnstable, Mass., about the slew of storms that hit Cape Cod in the winter of 2017. "We're like, what are we doing here?" he says, opening his arms skyward.

It is now the peak of summer as I stand with Matt in the seaside community of Blish Point at the front door of the house he owns—a house that's about to be demolished. Matt, 43, with a trim graying beard and a belt buckle in the shape of a fishhook, is the owner of REEF Design & Build, which works all across Cape Cod. He bought the house with his brother and father more than 10 years ago as an investment. Blish Point, an area where native fishermen once laid out their nets to dry, today contains a couple hundred homes nestled between the mouth of Barnstable Harbor and the verdant marsh of Maraspin Creek. Some of the homes are upscale; others are simple cottages. The Teague house, one of the simple cottages, was ruined by flooding: five major storms in the past three years alone have struck this area, and two of the four nor'easters last winter inundated the ground-level home.

Matt pushes his sunglasses atop his head, revealing a pale strip of untanned skin along his temple, as he stretches out his hand 2 feet above the door's threshold to show me where the water rose to during the storms. Over his shoulder, a hungry excavator sits ready to begin its work as Matt's extended family arrives, setting up lawn chairs across the street from the doomed house, joking about who forgot the popcorn. They have come to watch the carnage.

In spite of his own rhetorical question, after the demolition, Matt is going to rebuild—not elsewhere, but right here, only higher.

"The new top of the foundation will probably be about here," he says, shifting his hand to 3 feet above the flood mark, indicating a spot level with the door knocker, about shoulder-height. REEF Design & Build is raising the foundation not only for Matt's replacement house, but for many of the coastal houses the company is already building. Construction companies like his are responding to local building regulations, which are in turn responding to the most recent flood maps issued by the Federal Emergency Management Agency (FEMA).

While Matt the small-business owner is earning a good living from the building boom, there's also Matt the boy-turned-man who has been connected to the harbor across the street all his life and doesn't want to leave. "My grandfather worked here. I work here. My son works on a fishing boat here. It's a pretty special place for us." Like many Cape Codders, he is not ready to leave, no matter how much the sea encroaches.

"It's the new normal, unfortunately," Matt tells me with a shrug. His plan to move up instead of away—an action at once complacent and defiant—is emblematic of the stubborn New England hardiness that has been part of the Blish Point community since it was settled by Europeans centuries ago. But the "new normal" keeps shifting, and flood maps are drawn from the wrack lines (high-water marks) of the past, completely underestimating the likely impact of most climate projections. It's a gamble to live on America's shorelines in the

21st century, and those who dig in as they build upward are hoping they can stay one step ahead of sea level rise and the storm surges that increasingly threaten their homes.

Red Blood Among the Blues

Although Massachusetts is a solidly blue state as a whole, it is pockmarked with red regions, including chunks of the Cape. At election time, lawns along scenic Cape Cod byways sprout signs supporting deeply conservative candidates who make nary a mention of climate change in their campaigns. Before the September primaries, signs plaster every intersection in support of the re-election of Barnstable County Commissioner Ron Beaty Jr. —a strong Donald Trump supporter who served time in federal prison for threatening to kill both President George H.W. Bush and Sen. Edward Kennedy.

Matt Teague is vastly more moderate than this—though he comes from conservative stock. His father, Edward B. Teague III, represented Barnstable in the state legislature for eight years in the 1990s, eventually rising to House Republican leader, and spent some time as a conservative talk radio host. Matt was a lifelong Republican himself—until about 10 years ago. Then, exasperated by how hyper-partisan American politics had become, he "walked away from all of it," as he puts it. But not all of it, exactly; he voted for Donald Trump in 2016. "I got to make a living, and I don't like handing my money out to other people," he says by way of explanation.

Matt has two top political concerns: low taxes and what will become of his beloved Cape. "What's that harbor gonna look like in 10 years?" he asks wistfully, taking a momentary breather from his usual fast-paced mode of talk. "I want it to be clean, and I want it there for my kids." While he generally trusts science, he stops just short of accepting the scientific consensus on climate change. Perhaps, he suggests, the recent rounds of flooding were just a statistical anomaly. He sees the shifting sands in the harbor and can imagine it filling up in a century and becoming a marsh as easily as he can envision all of Blish Point underwater. "Who knows what this will look like?" he says.

What he does know is that he loves the Cape; It "is my whole life."

Managed Retreat from the Shore? Not Here.

Sea level has been rising since the end of the last Ice Age, around 20,000 years ago, when Cape Cod was formed as the Laurentide Ice Sheet retreated. But the pace of sea level rise in the last century far exceeds the previous incremental increase that took place over eons. As land ice melts at the poles and warm ocean waters expand, sea level is rising at an accelerated rate along the Mid-Atlantic coast, from Cape Hatteras to north of Boston. At the same time, in a kind of double whammy, the land is sinking from natural geological processes. If greenhouse gas emissions stay at their current levels, New England could experience seas that are nearly 7 feet higher than they are today by the end of the century, according to state documents.

Homeowners can easily dismiss the grave risks of the nearby Pilgrim Nuclear Power Plant, one of the country's worst-rated nuclear plants, which sits directly on the edge of Cape Cod Bay. It's harder for them to ignore the tenuous future (and value) of their Cape homes when storm damage from sea level rise increases. A report released this year by the Union of Concerned Scientists showed that nearly 90,000 Massachusetts homes, valued today at $63 billion, could be at risk by the end of the century—not just during storms but chronically, dozens of times each year. This past summer, low-lying areas were inundated during new- and full-moon tides, leading to the strange situation of having flood advisories when there was no rain. According to the Union of Concerned Scientists report, more than half of the homes in Blish Point are at risk of going the way of Matt's cottage, needing to be raised or risk ruin.

In some parts of the country facing this scenario, communities are opting for "managed retreat," in which homeowners in vulnerable neighborhoods allow themselves to be bought out all at once by city and state governments, instead of having the government bail them out repeatedly. But seaside properties maintain their special allure on old Cape Cod, which has a flat sandy seascape with an unavoidably low-lying geography—and Barnstable's Blish Point is not entertaining such a radical idea as managed retreat. Although the state recently passed legislation allocating $2.4 billion toward climate change adaptation and other environmental protections across Massachusetts, of which the Blish Point neighborhood will receive $1.3 million, buyouts are not a priority.

Instead, in the wake of the floods last winter, it wasn't just Matt Teague, but many Blish Point homeowners who were wrestling with hard decisions as they asked the question: "What are we doing here?" Some, like the Teagues, made plans to demolish and rebuild. Some had had enough, and a flurry of "For Sale" signs appeared by the time the summer tourists arrived. Some formed a citizen group that was partly responsible for getting that $1.3 million in state funding, which they hope will go toward restoring the marsh so it can better absorb flood impacts, and ensuring safe exit routes when it can't. Meanwhile, the day after the wicked January 4 storm, the neighborhood children ice skated with delight on the flood waters that had frozen, glasslike, in the front yard of Matt Teague's neighbor— while the same ice seized the interior contents of the houses and transformed them into wreckage.

Appreciating 'Common Sense Regulation'

Six weeks after the July demolition of the Teague residence, I step off the seventh rung of an extension ladder onto the new ground floor of Matt's freshly framed house, which is damp after a recent downpour. The plywood we're standing on is 13 feet above sea level, one foot above the minimum 12-foot base flood elevation height the town requires for this particular spot based on FEMA floodplain maps. Matt hopes to be the builder for nearly two dozen other houses in Blish Point, as they rise up, one by one, a town of extremely elevated homes defining Cape Cod's new normal.

Matt's cousin, Ian O'Connell, is there, too, and we all look down at the neighborhood from the new bird's-eye view. Ian, 40, has a big grin and curly dark hair peeking out from below a baseball cap with a fish logo. He missed the 2016 election, working on boats in the Caribbean and feeling like his vote wouldn't count in Massachusetts anyway, but he speaks positively of Trump's efforts to boost the economy. Like his cousin, Ian is fed up with both political parties; he considers himself an independent.

Ian is a service manager across the street at Millway Marina, which is owned by his father-in-law. During one of last year's storms, he used the snow plow on the front of his truck to part flood waters so he could get to work. Ian scrolls through his phone searching for photos of the storm's damage. One image shows a fuel pump half-submerged. Another shows a rack of boats. "See how close that boat is to floating?" he asks. "It's within two inches of lifting off the stands. ... It was scary." At one point, he says, the waters rose so high that a floating dock threatened to rise up off the piling that secured it. Ian had wrangled into waders and scrambled up the piling to bolt a vertical extension post, in the middle of the storm. It was New England can-do spirit in action.

I ask the two men what they think about how climate change, regulations and science all fit into this picture.

"We're on a warming trend," Ian says. "We're coming out of an ice age going towards a hot time. Is it happening faster than it should? I couldn't tell you." But he thinks it's ridiculous for politicians to take sides on the issue. "Leave it to a scientist to tell you that."

Matt keeps reverting to practicalities as a builder. He is grateful for state and local building codes that get stronger with each iteration. He praises officials for adopting "common sense regulations based on fundamentally real and good science." Even though he's not wholly convinced that humans are causing climate change, he says someone should pay attention to the data. "That is the role of the government." Despite what many of his fellow Trump voters might think, regulations "don't suck," says Matt. "They allow me to go to work." Of course, if a builder is required to construct stronger, taller—and more expensive—buildings because of climate change impacts, he can pass along costs to customers more easily than, say, Midwestern farmers facing extreme weather who rarely have control over what their crops and livestock earn them at market.

Ian isn't opposed to regulation, either. For those who live near the water, he says, regulations have had some positive effects.

"Remember the sheen that used to be in this harbor when we were kids?" Matt asks, referring to the iridescent layer caused by the oil and gasoline that spewed from inefficient two-stroke boat engines.

Ian brightens, that grin emerging again; he does remember it. "Everything has gotten better," he says. "Engines got better. Policy has gotten better. Groundwater testing has gotten better. Everyone's more conscious. Everything is more efficient and economical."

But while Matt and Ian have gotten used to the benefits from past regulations, they seem unconcerned about the possible dismantling of environmental restrictions that helped their community and beloved harbor thrive.

"They're not going to say, 'Dump your waste oil in the harbor,'" Matt says.

"'Let's bring the two-strokes back,'" Ian says, laughing. "That's not going to happen!"

But the Trump Administration is rapidly eroding environmental and climate protections. One week after our conversation in the shell of Matt's house, the EPA announced it would no longer require that coal-fired power plant upgrades include pollution-controlling scrubbers. It was one of dozens of rollbacks already secured or underway that could move the country farther from its Paris climate accord commitments and the Obama-era Clean Power Plan and closer to making 7-foot sea level rise a reality—and one that would arrive on Cape Cod's shores even sooner than originally projected, along with more powerful storms.

Salty Dogs and a Dose of Reality

Ian tells me to go talk to his father-in-law, Jack Hill, the owner of Millway Marina, if I want an even longer-term perspective of what's happened along the shoreline. I find Jack sitting quietly at his desk overlooking Barnstable Harbor. Now in his early 70s, he first worked this same harbor's edge when he was a teenager. Then, trash fish from the harbor were sent to feed mink on farms in Wisconsin, he says, and the tide didn't come as high as it does now.

Late-summer late-afternoon sunlight spills through the vertical blinds, across Jack's pale aqua eyes and teal shirt, down to the fading cornflower carpet, all echoes of the world of water that surrounds us. Beyond the window, past a couch with a life-sized plush toy black lab, I can see the extension to the piling that Ian banged into place last winter, mid-storm.

Jack is a man both steadfast and simmering. He tells me he voted for Trump. As a small-business owner, he says, his vote was driven by a desire for low taxes, and Republicans offered him more along those lines than Democrats. But about voting for Trump, he now says, "I'm sorry that I did, because he's a moron."

And when it comes to the issue of climate change, Jack thinks everyone in office is inept.

"All these politicians with their bullshit," he says, shaking his head. "There is a climate problem, and what are they doing about it? Nothing." I ask what he thinks they should be doing about it. Talking, he says. "You can't just completely say to one side, 'You're full of shit,' because then you're going to get nowhere."

The salt of the harbor flavors his language when discussing politics, but when a woman pushes open the squeaky door, asking about whale watch boat tickets, he directs her to the next building with a polite "ma'am."

"I just can't believe that straight, clear-thinking, halfway intellectual people can't see that there's a climate control problem," Jack resumes, leaning back in his chair. He'd like to see a comprehensive plan to address the problem for the sake of his kids, but "what's the plan now? Burn more coal? What the hell is that? That doesn't make any sense at all."

"You just have to figure out where the curves cross," he says as we finish talking. "How much can you enjoy without ruining the environment? I'm sure there are people thinking about it, but they sure are awful quiet."

The Ocean Creeps In

Follow the course of the shifting sands from Jack's office at Millway Marina and Barnstable Harbor and you'll wash ashore in Dennis, another Cape Cod town which, under the new FEMA flood maps, saw the number of homes at risk nearly triple. There you'll find Dan Fortier, a town planner who's not being quiet as he tries to turn adaptive strategies into reality. He gets some of his guidance from documents such as the state's climate change adaptation report, which recommends dozens of specific strategies to face the changes ahead. Some read like a Hippocratic oath of shorelines, directing a "No Adverse Impact" approach to managing coastal lands, while others promote using future climate change projections instead of historical data to estimate sea level rise and flood zones. But when we spoke, Dan kept returning to the economic risk for a place whose "export industry is summer."

With one-third of the residential properties in town being in a flood zone, "the impacts of the next storm are always on my mind," says Dan, who's worked with the town for 18 years. "If we lost one-third of our property value, it would be disastrous ... the death of our economy." That's the bind coastal towns find themselves in. They want to keep their citizens safe, but they depend on the property taxes of the most vulnerable of properties, which also happen to be the most valuable. At least for now.

Dan doesn't question the impact of climate change on the Cape. "Just in the last two decades, we have a continual creeping in of the ocean," he tells me. "The ocean doesn't recede the way it used to. Water is just there more and more because of sea level rise."

As tides ebb and flow, so do the tourists. By Labor Day, most of the cars loaded down with kayaks and sunburnt families will have driven away over the bridges, and the local kids will head back to school—though an unprecedented heat wave will cause the first day of some schools to be cancelled. By then, the tall concrete foundation wall of Matt's house will be hidden behind white siding, with shingles covering the upper stories that tower over his closest neighbors. Around the Cape, high-up houses like Matt's will keep sprouting. The 2018 hurricane season so far has been quiet around the Cape, but farther south, the Carolinas are reeling from an estimated $1 billion in damage from storm surge and flooding from Hurricane Florence.

During one of my conversations with Matt, I ask what he'll do if the 1-foot-above-flood-plain level that he chose for his foundation's height proves insufficient. "I designed it so I can jack it up again!" he says. He laughs, then pauses, becoming more reflective. "People are adaptive. Humans have always figured out a way to live where they live," he says. Consider the desert. The Arctic. Coastal areas. "The problem in the past was that people had to learn the hard way." Losing homes to floods is pretty hard, but New Englanders are used to hardship and hard weather and cleaning up after storms. Now they're getting used to building their homes higher and higher, hoping to reach themselves out of harm's way—and keep the view.

"The fact that there's enough science out there to provide some predictability for that and to provide for some policy—that makes sense," Matt says returning to the hope for smart policy based on solid science. "I think that's as good as you're going to get."

GENERATION CLIMATE: CAN YOUNG EVANGELICALS CHANGE THE CLIMATE DEBATE?

For students at this top evangelical college, loving God means protecting creation. That includes dealing with the human sources of climate change.

WHEATON, Illinois — Diego Hernandez wasn't thinking much about climate change until last summer, when he was traveling with his family along the Gulf Coast in his home state of Texas, where his ancestors—cowboys and politicians, he said—reach back to the 1600s. His mother suggested they take the "scenic route" for that summer drive, Diego said, his fingers making air-quotes because there was nothing "scenic" about it. All he saw were oil refineries.

"At that moment," said 19-year-old Diego, who considers himself a libertarian, "the switch kind of flipped for me." Why are we putting refineries in this beautiful place? he thought. The impacts from Hurricane Harvey, which had hit Houston the previous August and had affected some of Diego's relatives, were also still lingering in his mind.

"I used to be like, oh, there's oil, go start drilling, you know, because of course it's all about the money, right?" he said, his voice tinged with sarcasm. But after that family outing, he began to ask questions—"What is it doing to our environment? How is it going to affect us in the next 10 to 50 years?"—and since then he's had climate change on his mind.

Diego is a clean-shaven, lifelong Christian wearing a cyan blue button-down and polished cowboy boots, and a sophomore at Wheaton College in Wheaton, Illinois, which has been called the Harvard of Christian schools. The entrance sign, framed by a glowing bed of zinnias in full bloom, pronounces the school's motto: "For Christ and His Kingdom." But while Diego has all the credentials of a true political conservative—president of Wheaton's Young Americans for Freedom chapter, a cabinet member of the College Republicans—he also finds himself genuinely baffled by the right's stance against acting on climate change.

While many evangelicals are preoccupied with the long-term state of human souls and the protection of the unborn, Diego and the other students I met at Wheaton are also considering other eternal implications and a broader definition of pro-life. They are concerned about the lifespan of climate pollutants that will last in the atmosphere for thousands of years, and about the lives of the poor and weak who are being disproportionately harmed by the effects of those greenhouse gases. While Diego was just shy of eligible voting age in the 2016 presidential election, he's old enough to vote now. He and other young evangelicals thought hard this year about the politicians on offer, the issues they stand for, and who deserved their votes.

What's an Evangelical to Do?

Evangelical Protestants—one in four American adults—are a political powerhouse. They are the single largest religious group in the nation, and they are nearly twice as likely to be Republican as Democrat. And while Baby Boomers are currently the strongest political voting bloc, that's only because the older you are, the more likely you are to vote.

The current crop of younger people—from Gen X to Millennials to the newly minted adults I met at Wheaton—are poised to dominate the eligible-voter body politic. They would definitively tip the voting scales—should they become engaged. There are signs they might be doing just that. From the Parkland school shooting victims to Millennial political candidates, the youth of America are speaking up. And, significantly, they accept the scientific consensus on climate change at a much higher rate than their elders.

This is true even of young evangelicals, as the existence of the Young Evangelicals for Climate Action (YECA) attests. YECA is a ministry of the Evangelical Environmental Network that aims to mobilize students, influence religious leaders and pressure lawmakers into passing legislation to address climate change. I met Diego at a climate change discussion event on campus that was organized by Chelsey Geisz, a Wheaton junior and a YECA climate leadership fellow.

From Colorado Springs, Colorado, Chelsey, 20, always loved nature, she told me as we sat together in a gazebo in Adams Park, near campus. She'd taken a few classes on sustainability at Wheaton, and last year spent time working at Eighth Day Farm in Holland, Michigan, where Christian volunteers have turned the dirt once trapped below strip mall pavement into garden plots to grow vegetables for the hungry. These experiences meant she was primed when she heard about YECA.

Though non-partisan, YECA is targeting conservatives, since that's where the facts of climate change have failed to lead to action. According to the organization, they've engaged more than 10,000 young evangelicals so far. Along with Chelsey, there are another half-dozen fellows at other schools across the country, helping to build the grassroots movement. The fellowship includes a summer training session that covers the science of climate change, as well as the socio-cultural and religious aspects of the issue. As a YECA fellow, Chelsey organizes campus events such as the session I attended in September and she serves as Wheaton's executive vice president of campus sustainability, a new position that YECA helped develop.

It can be tough to be an evangelical who cares about climate change, Chelsey said, "because the environmental activists don't trust you and the evangelicals hate you." Or they could hate you; she was quick to point out that the evangelicals she knows personally are generally tolerant of her views. "I'm not encountering anyone at Wheaton, even among my most conservative friends, who disagree with climate change," she told me. She's having some trouble with her father, though, who's troubled by her YECA work. He holds a Harvard law degree, works at a company that invests in resource-rich properties, and associates Chelsey's transformation into a "climate activist" with a liberal agenda he finds suspect. "For a man who has such well-reasoned opinions, I just feel like there's so much emotion for him that it's not about the science at all," she said.

As for liberals themselves, Chelsey said, some of them do treat evangelicals like her with some suspicion. After all, aren't evangelicals the ones who elected anti-environment Trump?

"I think there's some misunderstanding about what our faith compels us to do," she said as the sun set behind her, creating a halo around the edges of her auburn hair.

Praising Natural Systems

Sean Lyon is a recent Wheaton graduate who was also a YECA fellow while he was in school. He feels that he was born to love the natural world; his first word as an infant was "bird," after all, and flying creatures remain a passion he can't quite explain. While in school, he created his own interdisciplinary major of biology and business and spent significant time in Tanzania working with ECHO East Africa, a faith-based sustainable agriculture organization. He still lives in the town of Wheaton, easy commuting distance to Chicago, where he's volunteering at the Field Museum of Natural History.

Sean, 23, grew up in upstate New York, among "classic North American white evangelicals," where climate was not a concern and politics were conservative. But his love of the natural world shifted his perspective. He saw heaven on earth, and something worth saving, in every wingbeat he witnessed.

"Every ecosystem carries His creativity in it," Sean said, "and every species is a mark of His design." He had a thick brass bangle encircling his wrist, and blue eyes behind clear Lucite-rimmed glasses. Sean drew an analogy to his sister and grandparents, who are all artists. "So how would I treat the art that they created? If I love them, then I'm going to treat their art well. I'm not going to deface it. I'm not going to ignore it. I'm going to really honor it. And so when I see my God as having created everything that I'm interacting with, I want to honor it because that's a way that I can show my love for this Creator."

But God didn't just create singular works, Sean said; he created systems, natural systems that every living being relies on. He hoped that all Christians—no, he corrected himself, all faiths—would unite to protect those systems.

"That's my current prayer."

'Structural Sin'

Climate science isn't questioned at Wheaton College the way it often is in the wider evangelical community. The school is a brick-and-mortar rebuttal to the myth that science and religion must be at odds with each other. When Wheaton students step into their-state-of-the-art science building, for instance, they are greeted with signs stating that a "sound Biblical theology gives us a proper basis for scientific inquiry," and a display featuring locally excavated Perry the Mastodon, which carbon dating shows to be more than 13,000 years old.

The school is not alone in intertwining commitments to love God and protect the earth, often referred to as "creation care." The Cape Town Commitment, a global agreement between evangelical leaders from nearly 200 countries, includes acknowledgement of climate change and how it will hurt the world's poor (and it is required reading for Wheaton freshmen). Katharine Hayhoe, an atmospheric scientist at Texas Tech University and an evangelical, has been an outspoken advocate for climate action. And in addition to YECA, there are numerous groups active in this arena, including the Evangelical Climate Initiative, Climate Caretakers, Care of Creation and A Rocha.

In late 2015, the National Association of Evangelicals (NAE)—the biggest umbrella group of evangelicals in the country, representing 43 million Americans—issued a statement accepting climate change, acknowledging the human contribution to it and encouraging action. YECA's advocacy helped bring that statement, called "Loving the Least of These," into being. In it, NAE argues that Christians should be compelled to care about climate change as a matter of social justice, equating those without the resources to adapt to failed farming or dry wells or rising seas as the modern-day equivalents of the widows and orphans of Jesus's day.

When Chelsey reads the Bible, she hears this gospel of social justice, too.

"Instead of talking about climate change," she said of her work as a YECA fellow, "I talk about environmental justice. There's definitely a guilty complex, especially among the white evangelical community, about how complicit we've been, and apathetic. People really want to redeem that."

Chelsey's framing reveals that she is steeped in a liberal arts ethos friendly to intersectionality, the idea that humanity's ills, which disproportionately affect the most vulnerable, cannot be conquered until root causes are addressed. This perspective is shaping academic dialogue in both secular and faith-based schools.

But does fighting climate change detract from evangelism? Here there's a rift within the evangelical community. Should the emphasis be on saving souls or saving God's creation? And are the two really at odds?

"That's the Billy Graham evangelicalism," Chelsey said of the personal salvation perspective, referencing Wheaton's most famous alumnus. "It's your faith between you and Jesus." But the problem with that approach, she said, is that it doesn't force Christians to deal with larger systems of injustice. "The evangelical community is really limited when it comes to talking about systemic and structural sin rather than individual sin. Most of us have never heard about systemic racism and climate change in church," she said. Even as evangelical organizations embrace the need for action, the message isn't coming across from the pulpit. "These things never come up because they're apparently not gospel issues," Chelsey said, "But at Wheaton, we think they are."

For Sean, there's not one speck of conflict between his love of God and the gospel and his fierce desire to see action on climate change. They're complementary, he said.

"If you focus too much on only a personal relationship being the core tenet of your faith, then it means that you're more easily able to marginalize topics like human suffering, which in some cases is spurred by climate change," he said. "We are embodied creatures in this planet, so let's live like we are."

Could his concern for the climate be a threat to his faith? I asked him.

"Actually, I see more of a threat in the idea that we can divorce our lives on this earth and the lives of other people and the lives of other creatures from our life of faith," Sean said. Better to revel in God's love. "How much deeper and how much more beautiful is a way of loving Him that involves my whole being and the whole world around me rather than just simply the status of my soul?"

When Pro-Life Means Entire Lives

Abortion was the entry point into American politics for many evangelicals, after the Supreme Court affirmed abortion rights in Roe v. Wade in 1973. Before that, evangelicals were generally unconcerned about abortion rights, which had the uncontroversial support of Republicans; they were also generally disengaged from voting. Today, the single-issue anti-abortion preoccupation of many evangelicals, now considered a given by many political leaders, confounds some of the young evangelicals I met at Wheaton.

"If we say we're pro-life, we have to care for people who are experiencing incredible environmental degradation and so directly affected by climate change," Chelsey said. "If we're pro-life, that's a bigger issue to me than abortion."

Sean agreed. "So many people are now saying, okay, if you're going to be pro-life you have to be pro all-of-life, lifelong pro-life, which has primarily come up in the immigration debate. If you're pro-life, how can you be separating children from their parents?"

Diego sees it a little differently. "Abortion is definitely a deal-breaker for me," he said, even though he said he's not generally a one-issue voter. He echoed Sean and Chelsey to some degree, agreeing that "being pro-life doesn't just mean being pro-life to the baby at birth. It also means the life of the mother and the life of the baby after birth." But when he watched the 2016 presidential debates, he found himself agreeing with some of Hillary Clinton's points ... until he was appalled by what he saw as her "gung-ho" support of abortion rights. He decided he could just not get behind someone with those views.

Young evangelicals wrestle with these difficult choices in the voting booth, confronted with either/or candidates, unsure who will best represent their hopes for life on earth, all life, all of God's creation. Right now, anti-abortion rights Christians typically have only one party to get behind. And it's that party, represented in the White House, that is aggressively rolling back climate protections, from pulling out of the Paris climate accord to promoting coal.

Future Powerhouse at the Polls?

Diego, Chelsey and Sean are the future. This younger generation has grown up with the realities of climate change and political polarization since they were swinging on monkey bars, and they aren't hesitating to break rank with evangelical Baby Boomers on the issue. They remain faithful and politically conservative for the most part, but they are more concerned about a climate that they will have to live with much longer than those boomers heading into retirement. The shift aligns with a recent Pew poll that found that among Republicans, young adults were far less likely than their elders to support reliance on fossil fuels.

"Every one of the people who I've talked to who's come to my events and engaged in climate issues from a Christian perspective said, 'My parents don't agree with me,'" Sean told me.

But even with this clear shift toward accepting climate science among young Americans, the quandary for young evangelicals in the voting booth remains.

Sean, who said he couldn't in good conscience vote for either party, opted for Jill Stein in 2016.

Chelsey, as a busy freshman in 2016, followed in her father's footsteps and voted for Trump. Her father had been singularly focused on getting a Republican on the Supreme Court. Now, she hangs her head about the decision.

Diego, about to vote in his first election, grew up in a struggling, hard-working family in San Antonio. His father showed him how to mow lawns when he was six, he said. His mother would pick up her raggedy old Bible and tell Diego, "This is what you should base all of your beliefs and all your values on. It shouldn't be what you hear from someone on TV or C-SPAN or NPR."

Surveys show that the way people view climate change is determined more by political affiliation, along with race and ethnicity, than by religious affiliation. So while 81 percent of white evangelicals voted for Donald Trump, it's important to remember that about a quarter of the country's evangelicals are not white, and it is among minority groups that the evangelical community is growing. And on the issue of climate change, Diego's Latino background makes him part of the American demographic that is most concerned about climate change. He wonders whether his mother deliberately pushed for that "scenic route" to wake him up a little.

What are the choices for these faithful young? With church membership in decline and the Republican party in flux, how vocal these young people are could shape the future of the climate debate. If the Christian right wants to hold onto the next generation, getting right with the planet might prove as important as getting right with God.

Many concerned about the environment rally for more

evangelicals to understand climate change and embrace leadership positions on the issue. "It would be a milestone if you managed to take influential evangelists—preachers—to adopt the idea of global warming, and to preach it," Nobel Prize-winning economist Daniel Kahneman told the host of Hidden Brain, an NPR science show. "That would change things. It's not going to happen by presenting more evidence, that is clear."

And in the book *The Creation: An Appeal to Save Life on Earth*, renowned biologist E.O. Wilson wrote a long letter with the salutation, "Dear Pastor." It is an urgent, heartfelt plea. "We need your help. The Creation—living Nature—is in deep trouble. Scientists estimate that … half the species of plants and animals on Earth could be either gone or at least fated for early extinction by the end of the century. A full quarter will drop to this level during the next half century as a result of climate change alone."

These new sermons and stories are unlikely to come from older pastors and preachers, most of whom have become representatives of the Republican Party platform that doesn't want to even acknowledge that climate change is an issue to discuss, let alone embark on the massive undertaking necessary to begin to solve it. But for the young, who will live with the catastrophic predictions that worsen with each new iteration of the UN climate report, there are new stories emerging. They are conversion stories of a new sort, springing from dirt once buried under Midwestern parking lots and held aloft on the wings of Sean's beloved birds. Preachers and politicians seeking to keep the young religious right in their midst may need to leap past the quagmire of a questionable climate change debate and get right to the root of finding solutions for the generations that will be living into the long tomorrow of a warming planet.

IT'S 'GOING TO END WITH ME': THE FATE OF GULF FISHERIES IN A WARMING WORLD

As global warming changes the Texas coast and cheap food imports flood the country, the people who make their living off oysters and shrimp are disappearing.

In early December, shrimper and oysterman Scooter Machacek, whose family has been working the Gulf Coast of Texas for four generations, took his two-man crew out to harvest oysters in waters off Palacios *(puh-LA-shus)*, a small port a couple hours down the coast from Galveston. It had been a terrible day, he tells me; it took seven hours to gather just 13 hundred-pound sacks of oysters, which his crew quickly unloaded from his boat "Hloczek" onto the deck at JoJo's, their buyer in Turning Basin 4. Thirteen sacks was a pitiful haul, less than half of the allowable daily limit of 30 sacks, which itself was a fraction of the 140-sack limit allowed in the heyday of the 1980s.

Everything, it seemed, was shrinking, and Scooter thinks there might not be a future in shellfishing along this stretch of shoreline.

"This is just a dying industry, is what it is," says Scooter, 53.

He might be right. The sleepy ports along the Gulf Coast of Texas are shadows of what they were just 25 years ago. Then, thousands of smaller bay boats like Scooter's, around 40 feet long, plied the shallow waters around the barrier islands that form the fringed coast. Thousands more of the larger gulf boats, at least twice the size of bay boats with double outriggers that stretch out like wings, disappeared into the deeper, federally controlled waters of the Gulf of Mexico itself, at least nine nautical miles out to sea, for weeks on end.

At the time, only a smattering of the public was paying attention to scientists raising the alarm about the planet's temperature. Diesel fuel cost about one-third less than it does today. U.S. Immigration and Customs Enforcement didn't exist. And Mexicans moved with relative ease over the border to join the rest of the shellfishing workforce: Anglos who had inhabited Texas for generations along with recent Vietnamese immigrants who had gravitated to the southern shorelines of the United States after the American government laid out the welcome mat for them, feeling a moral obligation to Southern Vietnamese after the fall of Saigon.

Everything is different now. The decline of the American shellfishing industry is inextricably linked to global systems both economic and environmental. From cheap food imports to hurricanes fueled by a warming planet, these systems support, or strain, the tapestry of what it takes to get seafood to the dinner plates of diners.

But part of the shellfishing struggle is that it is also collateral damage of a success: the achievement of the American Dream, by generations of shrimpers and oystermen who've come before. That dream might mean the end of Scooter's way of life because those who worked the hardest—and the labor of shellfishing is unmistakably hard work—supported the education and pursuits of their children, who are increasingly finding other professions to enter. The fate of the boats that line the port of Palacios, self-declared "Shrimp Capital of Texas," hangs in the salty air.

Shrimp constitute the biggest haul along this part of the Texas coast, followed by blue crab and oyster, all of which begin their life cycles in the bays and estuaries that serve as the Gulf of Mexico's nurseries. But those ecosystems are changing.

According to the latest National Climate Assessment, the Texas coastline is particularly vulnerable to such climate catastrophes as flooding, drought, increasing storm intensity and sea level rise. These climate change effects can be seen as a "threat multiplier" as they play out in the lives of men like Scooter—bay shrimping from May to November and oystering through the winter—who struggle to keep an industry alive.

Consider, for example, the single factor of rain. Scooter has felt the impact of increasing heavy rainfall.

"We're getting drowned in freshwater," he says. "The rain is crazy. That weather just comes and dumps and dumps and dumps. You used to get these little rains, now it just pours on." This influx of rain leads to a domino effect that's being felt up and down the Gulf coast.

Scooter's sure he'll be one of the downed dominoes. The long line of the family business, he says, it is "going to end with me."

Too Much 'Agua Dulce'

A couple hours up the coast from Palacios is San Leon, a peninsula jutting out into the Galveston Bay. It is an oystering hotspot. At Misho's Oyster Co., white oyster boats with bright blue trim come into port one after the other. These boats have met the 30-bag daily limit, though the oystermen tell me it took them seven hours when it should've been more like two. Each boat delivers its load directly onto a conveyor belt that carries the burlap sacks up to pallets that are loaded straight into trucks for distribution. The trucks then hit the road, some heading as far away as Virginia.

Once the boats have unloaded and are settled into their slips, I step from deck to deck of the tightly nestled vessels and hear variations on the same story from just about every captain. Too much "agua dulce," say Capt. Jose Tobar of the "Esmerelda," and Capt. Perez Martinez of the "Miss Joyce," and Capt. Jesus Delgado of "Buster." Freshwater is the biggest hindrance to the health of the oyster industry, they tell me, along with over-harvesting (as indicated by undersized oysters).

The Lone Star is a large state, and heavy rains in just about any river drainage can eventually be felt along the Gulf Coast. Excess freshwater comes from too much rain—or, as John Nielsen-Gammon, Texas's state climatologist, puts it, "the same amount of rain, but in more concentrated episodes." In every region of the United States, extreme precipitation events—when it just "dumps and dumps," as Scooter said—are rising along with human carbon emissions. If we can radically reel in emissions by century's end, they'll merely go up by half or maybe double compared to historical averages. If we don't, Americans could get "dumped" up to three times as frequently as they once did.

It makes annual rainfall averages misleading. The Texas coast might still be getting around 45 inches of rain each year. But for oysters—and humans—a deluge can be destructive in ways that steady rains are not.

The arrival of freshwater can initially be beneficial, bringing nutrients that nourish coastal ecosystems. But too much can easily wreak havoc. Oysters are especially vulnerable. Contained within their rock-hard shells, oysters are powerful filtration systems, but they depend on a certain level of salinity to function. When too much freshwater enters the system, the salinity plummets. The oysters can die, the shells yawning open, empty of life. When this happens, the Texas Department of State Health Services shuts down the affected harvest areas, forcing oyster boats to travel farther along the coast in search of open beds, or else to lie idle in port.

Heavy springtime rains can also affect the shrimping industry.

They can flush juvenile shrimp out to deep waters prematurely, where they're vulnerable to predation, because, as one biologist explained, "Everything eats shrimp." Flooding also brings a host of hazards in its wake. Storm debris can knock out ship propellers and tear up shrimp nets. Polluting effluents come from the region's prodigious petrochemical industries along with fertilizers from farm blitzes upstream. "When they spray the cotton to defoliate it," Scooter says, "and you get a rain after that, it wipes everything out, kills everything." And those who seek wild shrimp look askance at the growing number of shrimp farms, worried about exotic species and disease outbreaks.

In the past 30 years, floods actually came less frequently than in previous years to the southern Great Plains, a region that includes Kansas and Oklahoma along with Texas. But the floods that did arrive were repeatedly record-breaking. And as warming continues, climate models are predicting more of these heavy precipitation events across much of the southern and eastern United States.

Between the floods are droughts, when bay salinity shoots up. Oysters thrive in these conditions, but so do the parasites that attack them—again leading to oyster die-offs that cause the health department to shut down oyster beds. Extreme droughts have already resulted in losses to fish, crabs, oysters and waterfowl, and the swings between drought and flood are more common than a century ago. Efforts to artificially enhance declining oyster beds across the Gulf states are underway, including in Matagorda Bay in Palacios and in the waters off San Leon, where Misho's Oyster Co. dumps their used shells back into the water to help build future oyster reefs. Despite these efforts, though, if the water salinity and quality are poor, the oysters won't thrive.

With warming and drought conditions also come algae blooms. Already, red tides have become more frequent along the Gulf coastlines—they ravaged the coast of Florida earlier this year—and they are becoming more intense and more widespread, according to the National Climate Assessment. In ordinary circumstances, oysters can be miracle workers, filtering polluted waters yet themselves remaining clean enough for human consumption. But algae blooms are the exception. Even after a bloom clears up, toxicity can linger in the oysters for weeks or even months. Again, the health department shuts down harvests. In 2011, a bloom in Texas lasted from September into the next year, causing $7.5 million in losses to the fish and shellfish industries.

Shrimp Out of Sync

Even if heavy rains don't flush shrimp out of the bays prematurely, warming temperatures can spur their early departure. That would be fine if bay shrimpers didn't find themselves completely out of sync with the regulatory seasons.

"Usually we have brownies for a month," says Steve Pirhoda, referring to the brown shrimp that dominate the Texas harvest from May to June. "But this year, they were ready in April, a month before the season opened." Three days after he was allowed to harvest according to his state license, the brownies were gone to deeper waters, he says, beyond the reach of bay shrimpers like himself and Scooter Machacek, who is his cousin. He estimates that in 2018 he made half of what he usually does.

I meet Steve, 63, on the edge of Tres Palacios Bay in Palacios. An early December cold front is on its way and he zips up his Walls duck-hooded work jacket to fend off the wind. His shrimp boat "Sea Tiger" is lashed to pilings in the backyard of his waterfront home as he begins a gut renovation of the wheelhouse.

"Usually a hurricane helps us," says Steve, who has been

shrimping since 1974. Storms stir things up, which encourages feeding, so shrimping is often better in the year after a hurricane, despite the immediate disruptive aftermath. "But it didn't help us," he says about Hurricane Harvey, which struck this area of the Gulf of Mexico in late August 2017, causing more than a billion dollars worth of damage. "If anything, it got worse."

Steve dismisses climate change as a factor in what he's experiencing after more than 40 years of shrimping the same waters—he doesn't see the evidence, he says. He blames the decline on pollution from aquaculture, agriculture and a Colorado River diversion, as well as on the construction of an erosion-control barrier wall just beyond the marsh dunes that keeps young shrimp from moving with the tides.

Texas Parks & Wildlife (TP&W) has been sampling the Gulf for about the same amount of time that Steve has been shrimping, and their data match what he says about early departures. "We were able to predict within a week or so when you'd see juvenile shrimp" enter the Gulf, says Lance Robinson, deputy division director. He found that the shrimp would move from the bay into deeper waters "like clockwork"—but on a clock that had moved forward by at least a couple of weeks. In the 1980s and 1990s, the peak period of the shrimp moving into the Gulf was late May; now it's shifted to early May.

"Brown shrimp do appear to be leaving the bays earlier than they used to," TP&W science director Mark Fisher confirms, "most likely due to warmer water temperatures."

TP&W has recorded more than a 1-degree Celsius increase in temperature per decade in the 40 years of surveying. Agency officials are mulling over changing harvest dates, but they know there are still likely to be unseasonably cool springs in the future, which would delay the shrimp as much as warming speeds them up. So far, keeping the status quo of a May 15 opening day for the bay shrimping season is their compromise.

Same weather, same bay, same business, but Scooter Machecek diverges from his cousin on the matter of climate change.

"These storms are getting crazy," Scooter tells me. "They're getting bigger and bigger. And they develop quick, not like they used to. Ahhh, this is going to be a [Category 1 storm] and it comes in a 4! They think that's all it's going to be, and it just keeps a-ginning," whipping up in strength, "because the water is so warm."

The two men do agree on one thing: that the legacy of their livelihood might be coming to an end with them. Still, neither of them is quite ready to pack it up. "They told me 25 years ago that I wouldn't be shrimping in 20 years," Steve said. "We're laughing right to the end."

The Fading Fleets

There are essentially two fleets that pursue shrimp in these parts. The bay shrimpers, regulated by the state, stay close to shore, while the federally regulated Gulf shrimpers head to deeper waters. Both fleets are fading in size. State and federal license moratoriums and license buyback programs started in the 1990s and 2000s, prompted by a concern about overharvesting and a hope to maintain a viable livelihood for the boats that remained. That means that more threads of the already-fraying shellfishing industry have been clipped out of place. For the close-to-shore bay shrimpers, the state has bought back two-thirds of the licenses since the mid-1990s. Farther offshore, the number of boats has declined by a quarter just in the last decade.

On a bright December Monday I sit down with Craig Wallis, owner of W&W Dock & Ice, in his office in Turning Basin 2 in

Palacios. He has seven boats now, all of which were out on 45-day voyages in the Gulf. Each was expected to return the following week with 20,000 pounds of flash-frozen shrimp in their hulls. In the early 1990s, Craig tells me, he had twice as many boats, back when regulations were more lax and fuel was a lot cheaper. "That's a biggie for us," he says in his baritone voice, tapping his pen on the table. "Fuel—that's a third of our expense."

But while fuel prices have generally gone up (until the recent dip), prices for their catch at the dock have remained stagnant.

"Just like the beef industry or the chicken industry or anything else, nobody's getting paid any more for their product," Craig says. "There's prices we saw in the '70s that we see now," referencing the $2 to $3 per pound they get for their shrimp—a fraction of what consumers pay. "And expenses are going up, up, and up."

Even as seafood consumers become greater connoisseurs of what appears on their plates, they'll only pay so much for it, and by and large they don't really care where it comes from. That explains why 80 percent of the seafood Americans consume is imported, according to NOAA Fisheries. In 2017 shrimp imports alone were valued at $6.5 billion, and those numbers are on the rise. Those cheap imports are coming from shrimp farms in India, Indonesia, Ecuador, Vietnam and elsewhere, where labor and environmental laws are a shadowy sliver of U.S. standards. It's nearly impossible for American producers to compete.

Craig is 66, and his seven captains, all Latino and many of whom have been with W&W for decades, are all around the same age, that age of failing knees and bad backs. The industry relies heavily on younger H-2B visa workers from Mexico, but recent restrictions on this temporary non-agricultural laborer visa category, which includes everyone from waitresses to deckhands, have added a new challenge for an already struggling industry. Another thread. Another snip.

"We can't find local people that want to do that work," Craig says. "The caliber of white people in this particular industry," he continues, "they're not any good." He speaks highly of the Mexicans he's hired and worked with for decades. "They're here to work. They do a good job. They're high quality." As we speak, the wife of one the captains (herself an occasional deckhand, when labor's short), steps in with a Christmas gift for Craig, a nice bottle of Don Julio tequila. Of the 18 H-2B visa workers that W&W requested this year, they were permitted only two.

Infrastructure, Built and Economic

Another climate risk for Texas Gulf towns is the sea itself, which these days sometimes rises high enough to reach coastal properties with a kiss (if they're lucky) or a clobber (if they're not). Sea levels along the Texas coastline have gone up 5 to 17 inches, higher than the global average, in part because of subsidence caused by extraction of groundwater and fossil fuels. Along the Texas coastline, one thousand square miles of the Texas coast exists within five feet of the high-tide line, putting $9.6 billion of real estate, including hospitals, power plants and hazardous waste sites, and 45,000 people at risk.

Rising seas translate to more severe storm surges when hurricanes hit. One study found that global warming was responsible for making the rains from 2017's Hurricane Harvey about 15 percent more intense. Today, to reach the front door of some seaside homes, you first have to climb two flights of stairs.

The state of Texas is considering massive barrier-building projects to protect residents from the increasing threat of storm surges. Working with the U.S. Army Corps of Engineers, the state is conducting the Coastal Texas Study, a sweeping plan to create a more defensive shoreline that would include infrastructure such as dikes, levees and seawalls. How effective these forms of hard infrastructure would be is highly debatable—especially if inland flooding rushing down toward the Gulf is as much a threat as the storm surge coming up from it.

One obvious risk of these physical barriers is the one Steve Pirhoda points out, indicating the seawall that was built just beyond where the "Sea Tiger" is docked: the interruption of the necessary movement of species to and from the nurseries where their aquatic lives begin.

Unraveling

But there's another type of infrastructure at risk along the coast. Palacios was spared the wrath of Hurricane Harvey, but a bit farther down the coast, Port Lavaca was hammered. W&W Dock & Ice had shrimp being processed at a plant there whose roof caved in during the storm. They scrambled to send the shrimp south to a freezer that was functioning, but even after the storm was over, the plant never reopened. Now, W&W sends its shrimp on eighteen-wheelers to Louisiana for processing.

"If you start losing the infrastructure of a port, you can't get things done," Craig says. Welders, or the guys who do winch work, "they're out of business if they don't have enough boats to survive."

Just as the effects of climate change serve as a threat multiplier, there are economic multipliers, too. TP&W estimates that the close-to-shore bay shrimping industry, for example, brings in 1.8 times what the shrimp landings themselves earn the shrimpers. That means that the $8 million that ship captains received for their bounty in 2017 translated to more than $15 million for the local community.

The result is a shrinking and consolidation of the Gulf shrimping fleet. "In Port Aransas, there were 150 boats in the 1970s; now they've got two," Craig tells me. "Only three in Freeport, and they used to have a couple hundred." Many businesses folded, but others migrated to Palacios, keeping at least this port steady in the number of Gulf boats. All these places, "they've all diminished," Craig says, going as far back as the effects of Hurricane Katrina in 2005. "After Katrina, we lost a lot. Some weren't doing well so they took the insurance money and ran."

That was followed by more setbacks: Hurricane Rita just weeks after Katrina, Hurricane Ike in 2008, the BP oil spill in 2010. With each storm or corporate catastrophe, another snippet is clipped from the economic fabric of a place, and quietly it unravels.

The American Dream Endgame

Back in San Leon, not far from Misho's Oyster Co., lives Huynh Cong Tu, a longtime bay shrimper. He is a stocky spitfire of a man who was one of thousands of Vietnamese who fled from their country when the Communists took over. Huynh left his home on the Mekong Delta on a ferry boat loaded with refugees, made it to Malaysia, and from there he managed to get to the United States. It was 1979, the peak of the first wave of Vietnamese refugees, 130,000 looking for a new home after the war. He was 19 and eager to work, but without English skills, so he followed an uncle to the Gulf Coast and began working on boats, like many of his compatriots.

Many Vietnamese settled along the Gulf shores, from Alabama to Texas, and learned to work on boats. By the 1990s, some 60 percent of the Gulf fleet was Vietnamese. With the sudden demographic shift came violence; Huynh's uncle's boat was burned by the KKK in Seadrift, Texas. Huynh shifted to San Leon in 1982, when it was a virtually all-white town. Now San Leon is a melting pot: Latinos unload the oyster boats, a Nepali runs the local convenience store, and the red and yellow stripes of the Vietnamese flag fly from front yards around town. Each new group of immigrants comes, ready and hoping. Scooter Machacek and Steve Pirhoda's Czechoslovakian ancestors five generations back were no different.

"This is not easy money," Huynh tells me. "This is hard job. When we buy the boat, we have no money, but family help each other."

The Vietnamese community is tight-knit, pooling its resources and helping each other get established or get out of crises. After Katrina hit New Orleans, the Vietnamese communities recovered quickly because of these shared resources. "Borrow money from whole family. Put it all together, then one standing up. Then later on, the second one standing up," he says, a large pendant made of the tooth of a boar around his neck, the Buddha's figure carved within. "We try to do the circle like that."

But few of the owners of all those Vietnamese flags around town remain working on the water.

"Now, no shrimp," Huynh says. "Nothing. Less, less, less, all the time." He sold his boat and license in 1994 to take up crabbing instead, but the crab, too, seem to be disappearing. Over the last few years, he's seen the decline, sure there is something wrong with the water. But he's earned a good enough living to raise his house after Hurricane Ike swamped it, and to send all six of his children to school. Not one is considering following in their father's footsteps. They work as counselors and nurses, or go to school for law or medicine. Steve Pirhoda's son is heading into refrigeration. Scooter

Machacek's daughters are a nurse and a physical therapist.

"I got a couple grandboys," Scooter tells me, "but I hope they have nothing to do with it," speaking of his work on the water. "I hope they go get a real job, with real money. This ain't money. Not anymore."

It is possible that the shellfish will adapt to a changing climate, but the Americans that make their living off them will disappear anyway. All that unraveling, the dead yawning oysters, the shuttered processing plants, the sudden strength of storms, might make that coastal fringe fracture in a way it can't recover from.

When I ask Craig Wallis what the industry might look like in 10 years, he says, "There's going to be a commodity out there in the Gulf that you're not going to be able to afford to get."

But if you step back from the emptying docks, and pause, isn't this the American Dream in action? Anglo, Asian, Latino—everyone hopes for a better life, an easier life, a wealthier life, for their children. Take school seriously, stay away from drugs, cultivate a work ethic, and, in spite of staggering economic stratification in our society, this dream is still possible.

And the disappearance of shellfishing off our southern shores, maybe that's the collateral damage of the fact that Scooter's two daughters and Huynh's six kids can make a decent living that doesn't involve winches and weather. And when they're hungry they'll be able to afford nice dinners out with plates full of seafood, from somewhere.

ABOUT THE AUTHOR

Meera Subramanian is an award-winning journalist whose work has been published around the world. Her first book, *A River Runs Again: India's Natural World in Crisis from the Barren Cliffs of Rajasthan to the Farmlands of Karnataka*, was published in 2015 by PublicAffairs (and as *Elemental India: The Natural World at a Time of Crisis and Opportunity* by HarperCollins India). Her writing has been anthologized in *Best American Science and Nature Writing* and multiple editions of *Best Women's Travel Writing*. She was a Knight Science Journalism fellow at MIT (2016-17) and a Fulbright-Nehru Senior Research fellow in India (2013-14), and she earned her graduate degree in journalism from New York University.

Funding for this project was provided in part by a grant from the Roy A. Hunt Foundation.

ABOUT US

InsideClimate News is a Pulitzer Prize-winning non-profit, non-partisan news organization that provides essential reporting and analysis on climate, energy and the environment for the public and decision makers. We serve as watchdogs of government, industry and advocacy groups and hold them accountable for their policies and actions. Already one of the largest environmental newsrooms in the country, ICN is committed to establishing a permanent national reporting network, to training the next generation of journalists, and to strengthening the practice of environmental journalism.

We have grown from a founding staff of two to 19 full-time professional journalists and a growing network of contributors.

Climate and energy are defining issues of our time, yet most media outlets are financially unable to devote sufficient resources to environmental and investigative reporting. Our goal is to meet the growing need for coverage that can better inform members of the public in our democracy.

To help keep environmental journalism alive, donate to InsideClimate News by visiting:

https://insideclimatenews.org/about/membership

www.ingramcontent.com/pod-product-compliance
Lightning Source LLC
Chambersburg PA
CBHW070421220526
45466CB00004B/1501